Everyday
Mathematics
for Parents

Everyday
Mathematics
for Parents

WHO YOU NEED TO KNOW

TO HELP YOUR CHILD SUCCEED

THE UNIVERSITY OF CHICAGO

SCHOOL MATHEMATICS PROJECT

The
University of
Chicago Press
*Chicago and
London*

The University of Chicago Press, Chicago 60637
The University of Chicago Press, Ltd., London
© 2017 by The University of Chicago
All rights reserved. No part of this book may be used
or reproduced in any manner whatsoever without written
permission, except in the case of brief quotations in critical
articles and reviews. For more information, contact the
University of Chicago Press, 1427 E. 60th St., Chicago, IL 60637.

Published 2017

Printed in the United States of America

26 25 24 23 22 21 20 19 18 17   1 2 3 4 5

ISBN-13: 978-0-226-49375-6 (cloth)
ISBN-13: 978-0-226-26548-3 (paper)
ISBN-13: 978-0-226-26551-3 (e-book)
DOI: 10.7208/chicago/9780226265513.001.0001

Illustrations created by Bill Dickson,
represented by jupiterartists.com.

Library of Congress Cataloging-in-Publication Data

Names: University of Chicago. School Mathematics Project.
Title: Everyday mathematics for parents : what you need to know to help
    your child succeed / The University of Chicago School Mathematics
    Project.
Description: Chicago ; London : The University of Chicago Press, 2017. |
    Includes index.
Identifiers: LCCN 2016058314 | ISBN 9780226493756 (cloth : alk. paper) |
    ISBN 9780226265483 (pbk. : alk. paper) | ISBN 9780226265513 (e-book)
Subjects: LCSH: University of Chicago. School Mathematics Project.
    Everyday mathematics. | Mathematics—Study and teaching
    (Primary)—United States. | Mathematics—Study and teaching
    (Elementary)—United States. | Mathematics—Study and teaching—
    Parent participation.
Classification: LCC QA16.E84 2017 | DDC 372.70973—dc23 LC record
    available at https://lccn.loc.gov/2016058314

♾ This paper meets the requirements of
ANSI/NISO Z39.48-1992 (Permanence of Paper).

Contents

SECTION 1
Why *Everyday Mathematics*?

- Where are my child's math worksheets?
- Why does my child keep talking about math games?
- When will my child memorize the basic facts? Where are the timed tests and flash cards?
- Why does my child's homework jump from topic to topic every day? How is my child supposed to learn if the math keeps changing?
- Why isn't my child learning to add and subtract (or multiply and divide) the way that I did?
- How can I help my child with homework if it is so different from what I know how to do?

Does any of this sound familiar? If you are new to *Everyday Mathematics*, or just trying to understand it better, you are not alone. Many parents recognize that *Everyday Mathematics* is not how they learned math. They wonder whether their kids will learn this "new math." Or they worry that their kids will fall behind students at other schools. Knowing something about the background and philosophy of *Everyday Mathematics* will help you understand how it works and why it looks different from the way you may remember learning mathematics in school. In this section you will learn the many reasons why so many schools around the world have chosen to adopt *Everyday Mathematics* as their math curriculum.

Because It Is Based on Proven Principles

One reason why so many schools select *Everyday Mathematics* is because it is a pioneering program that has been developed, tested, and refined for over thirty years by teachers and researchers at the University of Chicago School Mathematics Project (UCSMP) to improve mathematics education across the United States. Since its launch in the mid-1980s, the program has established itself as the leading research-based elementary mathematics curriculum in the country.

Before embarking on the first edition of their new math curriculum, researchers at UCSMP undertook a worldwide study of how mathematics was being taught in schools. They translated and analyzed textbooks from top-performing countries and compared them to textbooks from the United States. In addition, the researchers wanted to understand exactly how children learn mathematics. They reviewed scholarly literature that was emerging, and they carried out their own research with teachers and children in classrooms.

In one early study, UCSMP authors surveyed incoming kindergartners to learn what they knew about mathematics at the start of the school year. They found that children come to school with far more mathematical knowledge than math curricula used at the time seemed to assume. For example, most textbooks at that time expected children to count and know numbers up to 20 by the *end* of kindergarten. UCSMP researchers found that 46% of children *entering* kindergarten could already count to 30. It was a startling finding. In fact, many children coming into kindergarten already knew much of the mathematics they were expected to learn during their first year of school.

Here are some other UCSMP findings:

- U.S. mathematics textbooks at that time focused almost exclusively on paper-and-pencil calculation. The idea of children learning how to use mathematical tools such as calculators was frowned upon.
- Most U.S. mathematics textbooks were organized to teach math skills in isolation, without linking the skills to each other or to the underlying mathematical concepts in a way that could make them interesting to children.
- Children actually learn mathematics better when they are given the opportunity to connect their everyday experiences to what they are learning. Learning to use mathematics, to apply it to everyday problems they understand, is particularly

important—and was largely neglected in textbooks of the time.

- And, contrary to widespread expectations at that time, all children can learn mathematics. Girls and underrepresented minorities can learn mathematics as well as anyone if they are given the chance.

This research shaped the first edition of *Everyday Mathematics* by helping the authors establish the following principles to guide them as they wrote a new curriculum:

Recognize that children begin school with a great deal of knowledge and intuition on which to build. Research has shown that children begin school with an intuitive knowledge of mathematics and abundant common sense. To be effective, a mathematics curriculum must meet children at their level and build on what they already know.

Connect children's learning of mathematics to their own experiences. A mathematics curriculum must begin with children's experiences and work to connect those experiences to mathematics. Fortunately, this is easy to do because basic mathematics is such a big part of daily life. Making connections to their everyday experiences gives children another reason for learning mathematics.

Emphasize excellent instruction. As with any subject, excellent instruction plays a critical role in children's success. A good textbook must include features that help teachers provide high-quality instruction and reflect on how they can improve the way they teach.

Work with children's problem-rich environment. Research shows that the school mathematics curriculum should incorporate challenging, interesting, real-world problems from children's environments in order to nurture higher-order thinking skills.

Use distributed practice to build skills. A well-designed program of routine practice spaced out over time helps children build mathematical skills. With solid mathematical skills, they become able to provide quick responses to simple problems. This, in turn, frees children to focus on more complex problems requiring higher-level thinking. *Continue learning from classroom testing and teacher feedback.* Once the *Everyday Mathematics* authors began writing the curriculum, they worked with teachers to test every lesson in classrooms.

A challenging, real-world problem taken from third-grade *Everyday Mathematics.* © McGraw-Hill Education.

They observed the lessons as they were taught, gathered feedback from their teacher partners, and used their observations and teachers' feedback to revise what they had written. The author team developed the curriculum one grade at a time to ensure coherence across grades.

All subsequent editions of *Everyday Mathematics* have been developed based on the same rigorous process, guided by principles drawn from the most current scientific research about educating children and continually tested in classrooms. The result is that *Everyday Mathematics* today embodies the same key features that distinguished it from the start:

- A balance among all areas of mathematics, including arithmetic, geometry, measurement, estimation, logical thinking, reasoning, graphing, relations, variables, explaining mathematical ideas, and making mathematical arguments.
- An emphasis on everyday uses of mathematics, which helps make mathematics matter to children by connecting it to their everyday lives.

- Ample use of drawings, diagrams, manipulatives, and other representations, which help children develop mathematical understandings and learn how to communicate their understandings to others.
- Opportunities for children to explore and discuss multiple strategies for solving problems. This teaches them to consider the correctness and efficiency of various strategies and to think about which strategy is most appropriate for a given problem or situation. And children learn from discussing the different approaches in class.
- Framing mathematical concepts in terms of children's problem-rich environment, which stimulates interest and encourages long-term learning and depth of knowledge.
- Use of continuous feedback, including classroom field testing, as part of the ongoing development process. In other words, authors design and draft activities and then pass them to teachers to be tested in their classrooms so that teacher feedback can be incorporated into revisions.
- Practice in the form of daily routines and games to make basic calculation skills fluid and quick.
- High expectations for all children, as well as the support needed to help children at all levels continue to grow in their mathematical understanding.

Everyday Mathematics is the result of a process that is purposeful and tested, designed to promote success across children's different learning styles and engage them in meaningful mathematics.

GO ONLINE
http://everydaymath.uchicago.edu/about/em-history

Because Spiraling Builds a Firm Mathematical Foundation

Another reason why *Everyday Mathematics* is such a widely used program is because of the "spiral." You may have heard that *Everyday Mathematics* is a curriculum that spirals. But what does it mean for a curriculum to *spiral*, and how is that different from other approaches? A spiral approach distributes learning over time. Topics are interwoven with other topics, and practice is spread out over time so that several skills and concepts are normally "in play" in any single lesson. Sometimes spiral curricula are also called *distributed* or *spaced*. Traditional mathematics programs are built differently: each topic is introduced, practiced, and completed in a single "chunk" of instruction. For example, a traditional math textbook may have a unit about geometry, another unit about fractions, another about probability, and so on. Once students have completed a unit, there is little or no opportunity to review or practice the mathematical topic, usually until the next school year. This approach is known as a *blocked* or *massed* approach.

So why does *Everyday Mathematics* spiral? The answer is simple: spiraling works. The research shows conclusively that spiraling generates better long-term mastery of skills and concepts and better ability to transfer what is learned in one context to other contexts. When measures such as end-of-year standardized assessments are used to evaluate long-term learning, children taught with a spiral curriculum like *Everyday Mathematics* outperform their peers taught using non-spiraling programs.

That spiraling works is clear. Why it works is less clear, and indeed spiraling strikes many people as counterintuitive. It seems more sensible to stick to one concept or skill until you've mastered it, and only then to move on to something else.

There are several reasons that might explain why spiraling is so effective. Distributed practice puts heavy demands on children's long-term memory, since they are frequently called upon to recall things they have not studied recently. These demands make the curriculum more challenging, but they also strengthen children's ability to recall from long-term memory what they have previously learned, with the effect that they learn it better. In addition, a spiral approach that revisits concepts over time in different contexts, like *Everyday Mathematics* does, also helps children make connections among mathematical ideas, deepening their understandings over time, reinforcing long-term learning, and improving their ability to transfer their learning to new contexts. And the act of recalling something from memory actually strengthens the ability to remember.

With massed learning and practice, on the other hand, children rely primarily on short-term memory. They may make quick improvements in performance in isolated areas, but studies show that their learning fades over time compared to children who first learn and then reencounter skills in a spiral. This may be because children taught with a traditional massed approach get less practice recalling what they know from long-term memory and have fewer opportunities to make connections across topics and build robust understandings.

You may be wondering whether spiraling can work for your child and his or her specific learning style or differences. Research shows that spiraling can been successful with all learners, including those with learning challenges. Because spiraling continuously exposes children to important skills and concepts, teachers are able to pinpoint learning difficulties early and can provide remediation the next time the skills and concepts turn up. In *Everyday Mathematics*, topics are revisited frequently, allowing all children ample opportunity to master them. It is important to note here that *revisiting* is not the same thing as merely *repeating*. Topics are treated in different ways and in new contexts, allowing children to extend understandings and preventing boredom, even in children who easily mastered the topic at the first introduction.

Since spiraling is so powerful for learning, you might wonder why other math programs don't use it. We mentioned one reason already: Spiraling is counterintuitive to many people; it seems illogical. Often teachers and students believe that a massed approach leads to higher performance. Did you ever cram for a test when you were in school? You may have passed the test and may have concluded that cramming (or massed practice) is a good way to learn. But chances are that not long after the test you forgot what you learned by cramming. Despite the feeling that cramming helped you learn, research shows that the learning was unlikely to be long-lasting. Robert Bjork, a psychologist at UCLA who studies distributed practice and other "desirable difficulties," has a term for the feeling you get from a massed approach: the "illusion of competence." You may feel as though you have mastered the material, but the "mastery" is likely to fade quickly. Your "competence" is not real and long-lasting; it's an illusion that will soon disappear.

Another reason spiraling is not common may be that teachers don't see the long-term negative effects of a massed approach. Teachers who use a massed approach may see their students perform well on end-of-unit assessments, but because the topics are not revisited or assessed again later in that year, they may not realize how much their students forget. That leaves it to the teachers in the next grade to see the effects most clearly, puzzling over why children in their class have no memory of learning a skill that was taught the prior year.

A final reason that spiral curricula are not common is that building a spiral curriculum is hard, and many textbook publishers lack the time, staff, and expertise to accomplish it. *Everyday Mathematics* works because of complex and sophisticated arrangements of instruction, practice, and assessment within units, within grades, and across grade levels. Designing and building a curriculum that spirals is much more difficult and time-consuming than designing and building a curriculum that takes a massed approach. Nevertheless, research shows that the benefits far outweigh the challenges.

FURTHER READING

Brown, P. C., Roediger, H. L., & McDaniel, M. A. (2014). *Make it stick.* Cambridge, MA: Harvard University Press.

Carey, B. (2014). *How we learn: The surprising truth about when, where and why it happens.* New York: Pan Macmillan.

Carpenter, S. K., Cepeda, N. J., Rohrer, D., Kang, S. H. K., & Pashler, H. (2012). Using spacing to enhance diverse forms of learning. *Educational Psychology Review, 24,* 369–378.

Cepeda, N. J., Pashler, H., Vul, E., Wixted, J. T., & Rohrer, D. (2006). Distributed practice in verbal recall tasks: A review and quantitative synthesis. *Psychological Bulletin, 132,* 354–380.

Delaney, P. F., Verkoeijen, P. P. J. L., & Spirgel, A. (2010). Spacing and testing effects: A deeply critical, lengthy, and at times discursive review of the literature. In B. H. Ross (Ed.), *Psychology of learning and motivation: Advances in research and theory*, Vol. 53 (pp. 63–147). New York: Elsevier.

Dempster, F. N. (1989). Spacing effects and their implications for theory and practice. *Educational Psychology Review, 1*, 309–330.

Dempster, F. N. (1997). Distributing and managing the conditions of encoding and practice. In E. L. Bjork & R. A. Bjork (Eds.), *Human memory* (pp. 197–236). San Diego, CA: Academic Press.

Roediger, H. L., III, & Karpicke, J. D. (2006). The power of testing memory: Basic research and implications for educational practice. *Psychological Science, 1*, 181–210.

Because of What We Know about How Children Learn

You may have been helping your child with *Everyday Mathematics* and found yourself thinking, "This is *not* the way I learned math!" *Everyday Mathematics* probably *is* different from the way math was taught when you were in elementary school, and it may also be different from the way children you know are being taught math in other schools. *Everyday Mathematics* is built on research about how children learn, and many of its key features capitalize on that research. This is yet another reason why *Everyday Mathematics* remains such a popular mathematics curriculum. Knowing what those key features are and the reasons behind those differences will help you understand it better. This section contrasts key features of *Everyday Mathematics* with the more "traditional" approaches used by many other math curricula.

WHY *SHOULD* MATHEMATICS BE TAUGHT DIFFERENTLY?

Chances are you've heard adults say "I was never good at math" or "Math never made sense to me." Yet rarely do you hear someone declare, "I was never good at reading." Why is there such a difference? Mathematics has long been perceived as a subject that can't be understood by everyone or as one that is not applicable in the "real world." Do you remember sitting in math class wondering, "When will I ever use this?" Many adults' negative opinions, even fears, of mathematics can be traced back to the way they were taught, instruction focusing on memorization and speedy recall of facts, formulas, and steps to compute with larger numbers. Children who could easily memorize and reproduce information, whether or not they understood the meaning behind it, could be successful in math class. But mathematics is so much more than a collection of memorized formulas. Consider the following experience shared by an *Everyday Mathematics* parent:

> As a kid, I was always able to do well with math by following the rules and procedures my teachers showed me. I could "borrow" and "carry" really easily, but I never understood how it worked. When my son started learning different ways to add and subtract bigger numbers in his second grade *Everyday Mathematics* class, I was impressed. The ways he learned actually made sense to me and fit much more with how I think about mathematics as an adult. Working on homework with him helped me to see how using different groupings of hundreds, tens, and ones can help with calculations. Now I finally understand the carrying and borrowing I was doing all those years ago.

Since the beginning, the authors of *Everyday Mathematics* have been committed to providing children with opportunities to learn mathematics in meaningful ways that develop their skills and understanding and boost their confidence in their abilities and their appreciation for the beauty, power, and utility of mathematics. Below we describe some key features of the curriculum that reflect this attempt to enrich learning for all students so that they enjoy, understand, and succeed with mathematics.

HOW THE STRUCTURE OF
EVERYDAY MATHEMATICS IS DIFFERENT

Do you remember spending much of fourth grade trying to master long division? Or never learning any geometry because it always came at the end of the book and your class never got that far? We've already discussed the way traditional programs tend to teach one skill or concept for a certain period of time before moving on to the next topic—never to return. Children taught using this massed approach may seem to "get it" when first assessed, but when topics are not revisited, learning fades quickly. As a result, a substantial chunk of the following school year is spent reteaching this forgotten material. Thus, the massed approach most mathematics textbooks take leads to a great deal of repetition every year and is a root cause of shallow and repetitive U.S. mathematics curricula. In contrast, *Everyday Mathematics* is structured so that children revisit skills and concepts periodically throughout the year, leading to enduring mastery and deeper understanding.

Everyday Mathematics also differs structurally from more traditional programs by building on what children know outside of mathematics, connecting mathematical ideas to everyday situations. Traditional programs tend to introduce mathematical terms and symbols at the beginning of a lesson, use them in a set of rote exercises, and then finish up by incorporating one or two "real-world" number stories at the end. For example, a traditional third-grade lesson on beginning multiplication would introduce the term "multiplication" and the symbol "×," present some strategies (such as skip counting or repeated addition) to solve multiplication problems, and then provide a set of simple multiplication exercises, such as 2 × 4 = ? and 5 × 3 = ?, for children to complete. *Everyday Mathematics* works differently. It introduces multiplication with a series of number stories like this one: *Ellie bought 3 packs of stickers. There are 6 stickers in each pack. How many stickers in all did Ellie buy?* Children draw pictures, use counters, or use other tools or representations to model and solve each story, and eventually they notice a common feature in their varied representations: in all of them they created equal groups. With this foundation in place, teachers are able to introduce the term "multiplication" as a way of describing problems that involve finding the number of objects in equal groups, thus connecting this new mathematical term to the intuitive notions children relied on when challenged to solve the problems in their own way.

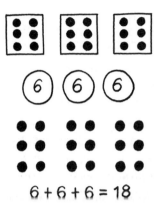

Children naturally form equal groups to make sense of a multiplication number story.

$6 + 6 + 6 = 18$

The learning environment in *Everyday Mathematics* classrooms is also very different from what is traditional. Do you remember your fifth-grade math class? Perhaps you watched your teacher work through several problems on the board before silently completing a worksheet full of similar problems. Contrast this with *Everyday Mathematics*, where collaboration, discussion, and problem solving form integral parts of the work children do and where students are sometimes expected to solve problems before they are shown how. By balancing teacher-led whole-class activities and discussions with small-group explorations and partner activities, along with some independent work, each lesson exposes children to a variety of ways to learn whatever mathematics may be involved.

Another structural feature that separates *Everyday Mathematics* from more traditional programs is the carefully designed set of student materials that work to encourage reflection, independence, and resourcefulness in young learners. These include a *Math Journal* for each grade and a *Student Reference Book* (or *My Reference Book* for grades 1 and 2) containing useful information keyed to individual topics. Traditional programs tend to use workbooks with tear-out pages in the lower grades and reusable textbooks in upper grades, requiring children to record answers on notebook paper. In both cases this means that children have no easy way to look back at and make use of previous work or to track their own progress.

Third-grade *Math Journal* and *Student Reference Book*. © McGraw-Hill Education.

Math Journal pages, on the other hand, are not only the "workbooks" in which students solve problems, but also serve as a year-long record of children's work because the pages are not torn out. Over the year, children accumulate a running record of daily work that they can use as an important mathematical resource. The *Student Reference Book* (or *My Reference Book*), a mathematics encyclopedia written for children, is also there for children to use whenever they need it to review a previously introduced topic or explore new topics as they arise. (See "Linking Math in School to Math at Home" in section 3 for more information on reference books.)

Everyday Mathematics is based on the recognition that children must develop their conceptual understanding of mathematical ideas in ways that make sense to them, using a variety of manipulatives, or "hands-on" tools, for learning, to facilitate the process. Some of the manipulatives used in *Everyday Mathematics* are very common, such as dice, dominos, and number cards. These are used frequently at all grade levels to randomly generate numbers for use in activities or games. For example, in early grades children roll dice to generate numbers to be added together to practice addition fact strategies.

A number of other tools that are common in mathematics classrooms but may be unfamiliar to parents are also incorporated into everyday lessons and routines. These tools include base-10 blocks and pattern blocks.

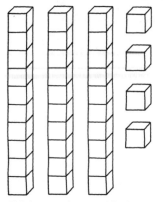

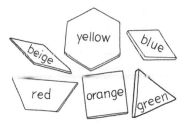

Children use base-10 blocks to learn about place value. In this image, the blocks represent the number 34.

A set of pattern blocks includes blocks of six different shapes, each shape a different color. Children use pattern blocks in many ways, including building figures and classifying shapes.

PRACTICE DOESN'T HAVE TO BE BORING . . .

Sometimes people mistakenly believe that because *Everyday Mathematics* makes such rich use of real-world problems, children don't have enough opportunities for practice. Nothing could be further from the truth. In fact, practice is woven into each and every *Everyday Mathematics* lesson in multiple ways and in a variety of formats. Practice is presented in small, frequent doses, reflecting the principle that the quality of practice is more important than the quantity. The practice routines include Math Boxes, games, Home Link Practice Strips, daily Mental Math and Fluency exercises, and Daily Routines. Most lessons begin with Mental Math and Fluency exercises—quick, leveled warm-up exercises that children answer orally, with gestures, or on slates or tablets, which encourage the development of mental arithmetic skills and number sense. Practice Strips, which are a handful of problems that appear at the bottom of selected *Home Links*, serve the same purposes.

Every lesson also includes a page of Math Boxes, a set of four to six problems that review previously learned content or provide a "sneak peek" at content yet to come. These practice sets integrate different areas of mathematics, such as geometry and number, so that children don't get "rusty" with content that has been previously taught and so they learn to recognize mathematics as a deeply interconnected subject. Math Boxes pages also include paired problems, which link practice in one lesson with practice of similar content in nearby lessons. The Math Boxes for Lessons 2-2 and 2-4, for example, will contain nearly identical problems, reinforcing key ideas as children progress from one related set to the next.

First-grade Math Boxes page.
© McGraw-Hill Education.

Finally, another major feature of *Everyday Mathematics* is the purposeful presence of games throughout the curriculum. To see how this works, consider the following situations:

> Classroom A: Second graders work quietly and independently, completing a worksheet containing 25 addition problems.
> Classroom B: Second graders work in partnerships, playing a card game called *Addition Top-It*. Partners take turns flipping over two cards from a stack and adding the numbers shown on the cards. They each say their answers aloud, and the child with the larger sum takes the four cards.

Children in both of these classrooms are practicing basic addition facts. Let's reflect on several benefits of the style of practice in classroom B as compared to classroom A. Classroom A illustrates a traditional practice routine still seen in many classrooms today. Children spend significant portions of time working alone solving repetitive exercises. In contrast, in classroom B, an *Everyday Mathematics* classroom, students are playing a game designed to develop basic fact fluency. Games provide opportunities for "meaningful practice," that is, practice that is not only fun, but also mathematically enriching. In fact, children often enjoy the games so much they don't realize they are practicing mathematics.

Another major advantage of this form of practice is the chance it offers for discussion. Because the children are invested in the game, they are motivated to question each other's answers, share their thinking, and help each other resolve errors as they play. This is more like life outside of school, where people talk and work together all the time. The traditional practice in which children complete math exercises independently means they lose the opportunity to reflect on and discuss what they are learning. And, quite honestly, it's rather boring. For all these reasons, *Everyday Mathematics* incorporates game play as an intentional and integral part of the program, both for practicing concepts and skills and for exploring new content. Occasionally, parents (and even teachers) may skip these games or replace them with rote practice because they do not realize the critical practice opportunities the games provide. Such a modification undermines a critical feature of the program and destroys its integrity. In *Everyday Mathematics*, games are *not* a supplement; they are an essential and meaningful part of practice and learning.

Because of How Key Content Is Taught

As you have been watching your child learn mathematics this year, you may have found yourself pondering questions like the following:

- Why isn't my third grader taking timed multiplication tests?
- Why isn't my kindergartner learning how to use a ruler?
- Why doesn't my second grader understand what I mean by "borrowing"?

- Is my first grader really learning algebra?
- How am I supposed to help my child when this isn't the way I learned math?

If you have asked yourself any of these questions, you are not alone. If fact, these are exactly the kinds of questions the authors of *Everyday Mathematics* asked themselves as they wrote each edition of the books. Every time a choice was made to deviate from the way math is traditionally taught, it was based on research findings and field testing, always with the goal of helping children gain a deeper understanding of the mathematics they are learning. The result is that *Everyday Mathematics* teaches some content differently from how it is taught in other mathematics programs. In section 2 of this book, "Exploring *Everyday Mathematics* Content," you will gain a better understanding of some of this mathematical content, the unique ways in which *Everyday Mathematics* approaches that content, and the reasons *Everyday Mathematics* is different. This unique, research-based content is yet another reason why *Everyday Mathematics* is selected by such an overwhelming number of schools and districts across the country and around the world.

Because It Works

We have shared a great deal of information about the history and philosophy of *Everyday Mathematics*, describing in particular how it differs from traditional math curricula and, most likely, from how you were taught mathematics. We hope you now have a better understanding of how your child is being taught mathematics and why so many school districts choose *Everyday Mathematics*. But you may still have that lingering question: Does *Everyday Mathematics* work?

One thing we know for sure is that results matter. This is why *Everyday Mathematics* has been the most widely used elementary mathematics textbook in the United States for over 20 years—because it produces results. You may be thinking, "Of course the authors of a math program are going to tell me that it works!" We at *Everyday Mathematics* can do more than just *claim* our program works; we can actually *provide evidence* that it works.

Independent research studies, as well as data from individual schools and districts from all over the country that are using *Everyday Mathematics* consistently show that the program helps children achieve. For example, *Everyday Mathematics* was the focus of a five-year longitudinal study carried out by researchers at Northwestern University. This study looked at results produced by *Everyday Mathematics* over a five-year period. The Northwestern researchers based their conclusions on teacher interviews, classroom observations, surveys, written tests, and a variety of other data. The study concluded that *Everyday Mathematics* students consistently outperform students using other programs.

Another large study, the Tri-State Achievement Study, compared the effects of mathematics programs on student performance on state-mandated standardized tests in Massachusetts, Illinois, and Washington. The records of more than 78,000 students were collected and analyzed. Half of the students had used *Everyday Mathematics* for at least two years, and the other half of the students came from comparison schools using other programs. Researchers found that *Everyday Mathematics* students consistently outperformed students in comparison schools. Results showed that the average scores of students in the *Everyday Mathematics* schools were significantly higher than the average scores of students in the matched comparison schools. These results held across all racial and income subgroups. The results also held across different state-mandated tests, including the Iowa Test of Basic Skills, and across topics ranging from computation, measurement, and geometry to algebra, problem solving, and making connections.

Data from specific schools tell the same story. 2015 data examined the top 25 highest performing school districts in Maine, based on district test scores. Of the 25 districts, 17 use *Everyday Mathematics*. Similarly, once students in the Denver Public Schools began using *Everyday Mathematics*, they made significant gains on the Transitional Colorado Assessment Program in all three grades tested.

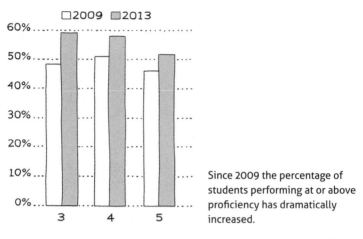

Denver Public School District Transitional Colorado Assessment Program—Mathematics Percentage Advanced or Proficient, Grades 3–5

□2009 ▨2013

Since 2009 the percentage of students performing at or above proficiency has dramatically increased.

Everyday Mathematics has been in use in schools in Horry County, South Carolina, since the 1996–1997 school year, when a pilot program in grades K–2 was introduced, with adoption expanding to grades 4 and 5 in 2000. *Everyday Mathematics* continues to be their curriculum of choice today. According to reports from the local Mathematics Learning Specialist, Horry County students leave school with more mathematical knowledge, a better understanding of mathematical concepts, and improved problem-solving skills, compared to students before the program was implemented. Students in Horry County have likewise had well-documented success with *Everyday Mathematics*, consistently outperforming peers across the state both in overall mathematics proficiency and in proficiency for each specific state mathematics standard.

Third graders using *Everyday Mathematics* in the Murfreesboro City School District in Tennessee in 2011 achieved an increase of 15% on the Grade 3 Tennessee Comprehensive Assessment Program. This was their first year using *Everyday Mathematics*. And in subsequent years Murfreesboro students have continued to widen the gap compared to their peers across the state.

Naturally, teachers and school administrators with such results

Palmetto Assessment of State Standards 2014 Math Standard 2
(varies for each grade) — Percent At or Above Proficient,
Grades 3-6

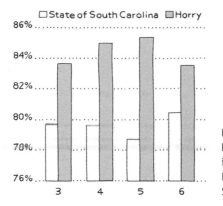

Horry County students scored higher than the rest of the state in each grade for their work in Number and Operations (Math Standard 2).

Murfreesboro City School District Tennessee Comprehensive Assessment Program (TCAP) — Mathematics Grade 3, Percent Advanced or Proficient

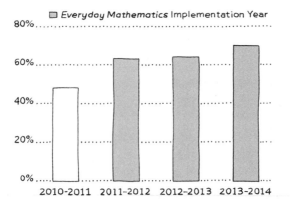

are pleased with the way their *Everyday Mathematics* students are performing. But what else are teaching professionals saying? Here's a sampling of what teachers have to say about *Everyday Mathematics*:

> "Fifth graders know so much more than they did in the past. They have become better mathematical thinkers and

can communicate mathematically both in writing and in discussion."

"I have become a better mathematician as a result of teaching *Everyday Mathematics* and feel strongly about how it teaches math and the transfer of concepts into the real world."

"The program connects the concrete to the abstract in a natural progression."

"I see team-building skills as students play the games. They actually reach more children."

"Students are empowered to share their strategies or methods of solving a problem, promoting diverse thinking."

These sentiments are echoed by administrators across the country. An Instructional Support Leader in Chicago, where *Everyday Mathematics* has been one of the approved math programs since 2003, says, "*Everyday Mathematics* makes sure that the students are focusing on what they need at every grade level, so that as they move through the grades, the curriculum is coherent."

The Abingdon School District in Pennsylvania has been using *Everyday Mathematics* since 1990. According to the Director of Curriculum: "I believe one of the outcomes of implementing the *Everyday Mathematics* curriculum is that it fosters good teaching."

In the Michigan City Area Schools in Indiana, users since 2003, the Director of K–12 Curriculum/Instruction says that students there "absolutely love *Everyday Mathematics* because of the hands-on approach and because there are multiple ways to get to an answer. But the bottom line is that *Everyday Mathematics* makes our students feel successful."

The Director of Teaching and Learning in the Edina Public Schools in Minnesota, users since 1989, keeps careful tabs on the numbers: "The *Everyday Mathematics* curriculum has students performing about one year ahead of traditional grade-level expectations."

Similar stories can be heard from districts across the country and around the world, in rural and inner-city settings, in large districts and in small private schools. *Everyday Mathematics* students outperform their non-*Everyday Mathematics* peers on standardized assessments and show dramatic gains year after year in all areas of mathematics. Perhaps most importantly, stories shared over the years by teachers and administrators consistently describe their students as actually *enjoying* learning mathematics!

So, does it work? The answer seems clear: *Everyday Mathematics* works. The program is effective in real classrooms with real students, and that's why, once schools have adopted *Everyday Mathematics*, they tend to continue using it, rolling over to each new edition as soon as it is published. As a result, many schools have been using *Everyday Mathematics* for more than 20 years.

FURTHER READING

Carroll, W. M. (2000). Invented computational procedures of students in a standards-based curriculum. *Journal of Mathematical Behavior, 18*(2), 111–121.

Carroll, W. M. (2000). *A longitudinal study of children in the "Everyday Mathematics" curriculum.* Chicago: The University of Chicago School Mathematics Project.

Carroll, W. M., Fuson, K. C., & Diamond, A. (2000). Use of student-constructed number stories in a reform-based curriculum. *Journal of Mathematical Behavior, 19*, 49–62.

Fuson, K., Carroll, W. M., & Drueck, J. V. (2000). Achievement results for second and third graders using the standards-based curriculum *Everyday Mathematics. Journal for Research in Mathematics Education, 31*(3), 277–295.

Sconiers, S., Isaacs, A., Higgins, T., McBride, J., & Kelso, K. R. (2003). *The ARC Center Tri-State Student Achievement Study.* Lexington, MA: The Consortium for Mathematics and Its Applications.

GO ONLINE
http://everydaymath.uchicago.edu/research

SECTION 2
Exploring *Everyday Mathematics* Content

While we don't have the space in this book to address in detail *all* the mathematical ideas your child will encounter throughout the program, rest assured that *Everyday Mathematics* does a thorough job of addressing *all* content children are expected to learn at each grade level, including a complete treatment of all content recommended in the Common Core State Standards for Mathematics.*

This section will focus on several key content areas: number stories, basic facts, operations with larger numbers, algebra, fractions, and early measurement. We chose these topics purposefully. Some topics, such as algebra, are important because, unlike *Everyday Mathematics*, in traditional programs they are not treated as a focus in the early grades. Other topics, such as facts and operations, deserve attention because they occupy such a large portion of the time children spend learning mathematics in elementary school.

Number Stories:
The Foundation for Success with Computation

Young children enter school with a wealth of real-world experiences that teachers can use to provide context for mathematical problems. By starting with children's own experiences, teachers can help them make sense of situations in which addition, subtraction, and even multiplication and division can be useful. Research shows that very young children are able to represent and solve real-world problems that adults would solve with addition

---

* For a brief grade-by-grade overview of the content addressed in *Everyday Mathematics*, see "What to Expect at Each Grade Level of *Everyday Mathematics*" in section 3. And, for a comprehensive listing of the Common Core State Standards for Mathematics, see www.corestandards.org/Math.

and subtraction even before learning terms such as "add" and "subtract" or symbols such as "+" and "−." In spite of this finding, traditional mathematics programs commonly begin teaching operations (addition, subtraction, multiplication, and division) with abstract equations like 2 + 3 = ? and 4 − 1 = ? They introduce "story problems," math problems set in real-world situations (known in *Everyday Mathematics* as "number stories"), much later, even though research suggests that story situations provide contexts for children that help them understand and apply the mathematical operations.

Everyday Mathematics differs from traditional programs in that it "flip-flops" the traditional sequence. *Everyday Mathematics* begins by providing context. Children are introduced to the operations using number stories that interest them and that they can explore, and, from these explorations, they develop their understanding of the operations. Thus, number stories provide children an intuitive foundation that all other work with operations can be built on. They learn not only how to add, subtract, multiply, and divide, but also how to make sense of the meaning of adding, subtracting, multiplying, and dividing, even from a very young age.

ADDITION AND SUBTRACTION NUMBER STORIES

In *Everyday Mathematics*, children begin their study of operations in kindergarten by exploring number stories with familiar contexts. Kindergartners act out the number stories or use counters or drawings to model and solve number stories before they even learn the words "add" or "subtract." Unlike other math programs, *Everyday Mathematics* is very deliberate about exposing children to all types of addition and subtraction situations (described later in this section) and even provides child-friendly names for the different kinds of stories. The result is that children leave the early grades of *Everyday Mathematics* comfortable with making sense of a wide variety of real-world addition and subtraction problems.

While the number of potential contexts for engaging number stories is essentially unlimited, simple addition and subtraction stories tend to fall into a limited number of categories. In the first category, the story involves some sort of action or some change that occurs. Some programs refer to these as "adding-to" and "taking-from" situations, but in *Everyday Mathematics* we call them *change* number stories. Consider these two change stories:

> Toto the terrier has 3 bones buried in her backyard. Today she buries 2 more bones. How many bones does Toto have buried in the yard now?
>
> Missy the poodle has 10 dog biscuits in her bowl. She eats 4 of them. How many dog biscuits does Missy have left?

The story about Toto is a change-to-more story because the starting quantity increases. To solve the problem, a child might begin with the 3 initial bones and count up 2 more to determine that now Toto has 5 bones buried in the yard. The story about Missy is a change-to-less story, as the starting quantity decreases. A child solving this problem might start with 10 counters (representing 10 dog biscuits), take away 4 counters, and count to see that there are 6 counters left. In both stories the unknown quantity comes at the end of the story. This type of story is typically easiest for children to solve because it is easy to act out or model using pictures, counters, or other tools.

In contrast, the following change-to-more story might be more challenging for most children:

> Ben has a collection of toy cars. His friend gives him 3 more cars. Now Ben has 12 toy cars. How many cars did Ben have to start with?

How does this problem differ from the prior ones? In this case, the starting value, or the number of cars Ben had at the beginning of the story, is unknown. Stories that begin with an unknown quantity can be much more challenging for children because it can be hard for them to figure out how to begin modeling the story, or solving the problem. Since the beginning amount is unknown, they struggle to get started. Many textbooks avoid number stories like this, or place them as optional challenges at the end of a page of problems. *Everyday Mathematics*, however, believes that it is important to give children experience with a wide variety of number stories and provides many challenging number stories for students to make sense of and solve. To help with these more challenging stories, *Everyday Mathematics* provides special diagrams (called *situation diagrams*) that children use to help them organize and make sense of the information given in a story. (For a more detailed discussion about modeling and solving number stories with situation diagrams and equations, refer to "Algebra in Kindergarten?")

The number stories children first encounter in *Everyday Mathematics* can typically be solved using either addition or subtraction. For example, in the story about Ben above, some children might think: "What do I need to add to 3 to get 12?" Then they might count up from 3 to 12 to determine that 9 is the correct answer. Others might recognize that they are trying to determine the difference between the two numbers, 3 and 12, and instead think of subtracting (12 take away 3 equals 9). It is helpful for parents to encourage their children to think flexibly about addition and subtraction in this way, allowing their children to use whichever operation makes sense to them in a particular situation, rather than prescribing one correct solution strategy.

A second type of addition/subtraction number story involves two or more smaller quantities, or parts, and a larger total quantity. Some programs refer to these as "putting-together" and "taking-apart" situations, but *Everyday Mathematics* has found that the label *parts-and-total* makes more sense to children. In a parts-and-total situation, the quantity being solved for can be either a part or the total, as shown in these examples:

> 3 purple flowers and 8 pink flowers are growing in a flower box. How many flowers are there in all?
>
> A vase holds 6 white flowers and some blue flowers. Altogether there are 11 flowers in the vase. How many of the flowers are blue?

In the flower box story, both parts (3 purple flowers and 8 pink flowers) are known and the total needs to be determined. The vase story provides only one part (6 white flowers) and the total (11 flowers). To solve the problem, children have to determine the missing part (the number of blue flowers).

In other parts-and-total situations, both parts are unknown. For example: *I have a bag of 10 blocks* (the total). *Some are yellow* (one part) *and some are blue* (another part). *How many of each color could I have?* Problems like this have many solutions and offer children valuable opportunities to practice breaking apart numbers in flexible ways.

Yellow blocks	Blue blocks
5	5
1	9
9	1
2	8
3	7
4	6
8	2
7	3
6	4

Solutions for the blocks parts-and-total problem

The third type of addition/subtraction situation involves *comparison*. In these problems, children are asked to compare two quantities, such as in the following problem:

Harrison read 4 books this week, and Thea read 9 books this week. How many more books did Thea read than Harrison?

To solve this problem, children need to compare the number of books Harrison read with the number of books Thea read. One strategy for solving this number story would be to count up from 4 to 9 to determine the difference of 5 books. Comparison problems are typically more difficult to solve than change or parts-and-total stories, in part because one of the quantities involved, the difference, is only virtual, not real. In the comparison of Thea's and Harrison's books, for example, the difference, 5 books, is an amount of books that Harrison didn't read or that he would need to read in order to have read as many books as Thea read.

Beginning in second grade, children begin to encounter multi-step number stories, such as this one:

> Miguel has 5 baseball cards in his folder and 8 in his backpack. Altogether, he has 4 more cards than Victor has. How many cards does Victor have?

These are called *multistep* number stories because it takes two or more steps to solve them, although there may be variations in the order of the steps leading to a solution. For this number story about baseball cards, children might first find the total number of cards Miguel has (13) and then subtract 4 to find out how many cards Victor has (9). You may have noticed that this story involves both a parts-and-total step (finding Miguel's total) and a comparison step (comparing the number of Miguel's cards with Victor's). In *Everyday Mathematics*, children begin by using drawings or situation diagrams to organize the information in these stories and eventually learn to model the situations using equations. Beginning in third grade, multistep number stories incorporate multiplication and division as well.

MULTIPLICATION AND DIVISION NUMBER STORIES

Everyday Mathematics introduces multiplication and division late in second grade. As with addition and subtraction, children's first encounters with these operations are contextualized in number stories based on familiar situations. Early multiplication and division number stories usually involve situations with equal groups or rectangular arrays of objects in rows and columns. Whether a particular number story is better solved using multiplication or division depends on what information you need to find. For example, consider these three number stories:

- You have 5 boxes of golf balls with 3 balls in each box. How many golf balls do you have in all?

- You have 15 golf balls that you need to share equally between 5 boxes. How many golf balls will go into each box?
- You have 15 golf balls that you need to put into boxes that hold 3 balls each. How many boxes can you fill?

The first story requires finding the product of 5 and 3 and represents a typical multiplication story. In the second story, both the total and the number of groups are known, so the problem is to determine the size of each group. *Everyday Mathematics* calls stories like this one *equal-sharing* stories, as the total quantity is "shared" between the known number of groups. The third story gives the total and the amount that goes into each group, and the problem is to determine the number of groups. *Everyday Mathematics* calls these *equal-grouping* stories. Both equal-sharing and equal-grouping stories involve division, but we can also think of them as multiplication stories in which a factor (that is, the number to be multiplied) is missing.

Just as with addition and subtraction, children begin solving these problems by using counters, drawing sketches, or acting out the stories to make sense of them. They might use counting, skip counting (for example, counting 2, 4, 6, 8), or adding up the groups to solve. Later, they also learn ways to model these situations using diagrams and equations (See "Algebra in Kindergarten?").

In *Everyday Mathematics*, children begin working with equal groups and array situations toward the end of second grade and continue working with these and equal-sharing problems in third grade. In fourth grade, children are introduced to another type of multiplication situation, called *multiplicative comparison*, in which children use multiplication to compare two quantities. Here is an example of a multiplicative comparison problem: *Helene checks out 7 books from the library. Ali checks out 4 times as many books as Helene. How many books does Ali check out?* Solving the problem requires children to compare the number of books Helene checked out with the number of books Ali has checked out—but to use multiplication rather than addition for the comparison. Children have been solving additive comparison problems such as *Helene checks out 7 books from the library. Ali checks out 4 more books than Helene. How many books does Ali check out?* for years, but multiplicative comparisons, which involve concepts such as "4 times as many" rather than simply "4 more," are a new kind of problem. Because the comparison is more complex, this type of problem is more appropriate for older children.

Children using *Everyday Mathematics* are first exposed to the four operations through number stories. This allows them to use familiar aspects of their worlds to model and make sense of the operations. Learning this way gives children a valuable foundation on which to build a more formal understanding of the operations. This also lays the groundwork for understanding and learning basic facts and strategies for computing with larger numbers. The next two sections provide more details about that important mathematical work.

FURTHER READING

Carpenter, T. P., Ansell, E., Franke, M. L., Fennema, E., & Weisbeck, L. (1993). Models of problem solving: A study of kindergarten children's problem-solving processes. *Journal for Research in Mathematics Education*, 24(5), 427–440.

Carpenter, T. P., Fennema, E., Franke, M. L., Levi, L., & Empson, S. B. (2015). *Children's mathematics: Cognitively guided instruction* (2nd ed.). Portsmouth, NH: Heinemann.

National Research Council. (2001). *Adding it up: Helping children learn mathematics*. J. Kilpatrick, J. Swafford & B. Findell (Eds.). Mathematics Learning Study Committee, Center for Education, Division of Behavioral and Social Sciences and Education. Washington, DC: National Academy Press.

There's Nothing "Basic" about Mastering Basic Facts

Mastering basic facts—single-digit combinations of numbers such as 7 + 5 or 4 × 8—is a major focus of mathematics learning in the early grades. Developing mastery of basic facts is critical for future success in mathematics because children use their fact knowledge when solving computation problems involving larger numbers. Yet there is really nothing "basic" for children about learning their basic math facts. It requires a great deal of time, well-designed instruction, and meaningful practice. Realizing this, the authors of *Everyday Mathematics* carefully designed a learning progression for facts spanning kindergarten through third grade. This approach, which includes a wide variety of practice activities, has been shown to help children graduate from the early elementary grades with lasting fact mastery.

If you have memories of learning your facts as a child, they are probably filled with flash cards, timed tests, worksheets, and lots of drills. Many mathematics programs continue to use such tools, in spite of the growing body of research that suggests that these tools are ineffective for *really* learning basic facts. *Everyday Mathematics* differs dramatically from other curricula in how it helps children learn and practice their facts. The program includes *no* timed tests, and the endless worksheets and drills have been replaced with games that are engaging, activities that are meaningful to children, and plenty of classroom and small-group discussions about strategies for solving facts problems.

However, just because the program doesn't use timed tests and worksheets doesn't mean it lacks opportunities for needed fact practice. On the contrary, opportunities for practice are built into the program, so that more time is spent practicing basic facts in first through third grade than on any other mathematics topic. The key for adults who went through a different learning experience is to recognize this sometimes unfamiliar, yet very rich, fact practice for what it is. Too often even teachers fail to recognize the way practice appears throughout *Everyday Mathematics*, and they replace the carefully constructed games and activities with more familiar forms of exercise, such as worksheets or timed tests they have picked up from outside the program. Doing so undermines the careful development of fact learning the authors of *Everyday Mathematics* intended. This section will help you better appreciate the meaningful and supportive ways *Everyday Mathematics* helps children develop understanding and knowledge of their basic facts.

When learning basic facts, children typically progress through three general phases for each of the four mathematical operations. In phase 1, they count using counters, pictures, or their fingers, and then counting. For example, to solve 4 + 5, children might count on their fingers to find the total 9. This is perfectly acceptable in phase 1. As they transition to using more efficient strategies to solve unknown facts, they move into phase 2.

In phase 2, children develop *fluency*, or the ability to *flexibly*, *efficiently*, *accurately*, and *appropriately* apply fact strategies. In other words, they learn to work from facts that they know to figure out facts that they don't know well. Thus, in phase 2, children might solve 4 + 5 by starting from an easier known fact, such as 4 + 4 – 8, and then adding 1 more to reach 9. This example demonstrates all four components of fluency:

- It is *appropriate* because the choice of strategy is reasonable for the given problem.
- It is *efficient* because the strategy allows them to find the answer easily and quickly.
- It is *flexible* because applying the strategy involves breaking the 5 apart into a 1 and a 4.
- And it is *accurate* because by carrying through the strategy, they find the correct answer.

As the result of having many phase 2 opportunities to practice facts, children gradually progress to *automaticity*—either automatically applying a highly efficient strategy or quickly recalling the answer from memory, so that the fact is solved within three seconds. Automaticity is the third and final phase of fact mastery, and children reach this phase with different facts at different times throughout the years.

Phase 2 is the most important part of this process, for that is when children have many opportunities to develop rich mathematical thinking. Children need ample time

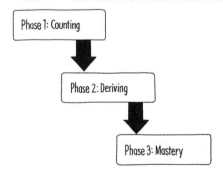

Phases of Mastering the Basic Facts

Phase 1: Counting

Phase 2: Deriving

Phase 3: Mastery

and support to create or make sense of appropriate strategies, to develop flexibility with applying those strategies, and, finally, to learn to apply strategies efficiently. Traditional math programs often lack opportunities for children to develop phase 2 thinking, instead forcing children to go directly from counting to rote memorization based on drilling with flash cards, worksheets, and timed tests. Often the result is short-term memorization that does *not* translate to long-term mastery. In other words, children initially appear to have mastered their facts because at the end of all the drilling, they successfully complete a timed math test. But after a week or a month, they no longer remember many facts and tend to fall back on counting as their only strategy.

In contrast, *Everyday Mathematics* helps children develop fact fluency by cultivating their understanding in phase 2, using a carefully developed sequence of lessons spanning multiple years. Children then move from fluency to lasting automaticity through frequent, varied, and purposeful instruction and practice.

DEVELOPING FLUENCY WITH ADDITION AND SUBTRACTION FACTS

We have already described how in *Everyday Mathematics* children first encounter the concepts of addition and subtraction through number stories in kindergarten. For example: *Toto the dog watched 4 squirrels playing in a tree and 2 squirrels burying acorns. How many squirrels did Toto watch in all?* Number stories such as this one use a real-life context to help children develop problem-solving skills and make sense of the meanings of addition and subtraction.

Other activities encourage children to develop flexibility in working with numbers, combining and taking them apart in various ways. For example, they learn to use connecting cubes of different colors to show different ways a given number can be broken into two parts. They might create 8 in many different ways, such as 4 red and 4 blue cubes, 5 red and 3 blue cubes, 6 red and 2 blue cubes, and so on. Children also learn to represent numbers in various ways using a tool called a *ten frame*—a table made of two columns with five squares in each column.

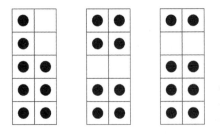

Ten frames showing various ways to represent the number 8. © McGraw-Hill Education.

Another way *Everyday Mathematics* encourages children to develop flexibility with numbers is by a routine called Quick Looks, which begins in kindergarten and continues through the end of third grade. In this routine, the teacher briefly displays a dot pattern or ten-frame image, so children register the image without seeing it long enough to count the dots. Next, the teacher hides the image, requiring children to use their mental imagery to try to make sense of the number of dots it contained. Finally, the class discusses the various ways different children saw the image.

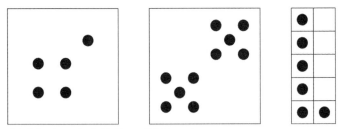

Quick Look cards. © McGraw-Hill Education.

The Quick Looks routine helps children develop the ability to *subitize*, or instantly see a quantity without having to count, and to flexibly decompose (or break apart) and compose (put together) quantities. By learning to do this with dots, young children develop the type of thinking that allows them to move beyond counting to creating more sophisticated and efficient strategies. Based on these early experiences, children begin to develop fluency with adding or subtracting 1 from a number (for example, 2 + 1 or 4 − 1) and with small number combinations such as 2 + 3.

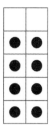

This Quick Look image encourages children to recognize the number 8 in any of four different ways.

- I saw 2 lines of 4, and I know 4 + 4 = 8.
- I saw 2 squares stacked up, and each square has 4 dots, making 8 total.
- I skip counted by 2s: 2, 4, 6, 8.
- I saw 2 were missing from the ten frame, so I thought 10 − 2 = 8.

Such exercises lay an important foundation for work with larger numbers. *Everyday Mathematics* builds on this foundation through numerous activities and specialized games that repeatedly expose children to two key groups of addition facts. These key groups of facts are the *doubles*, facts for which both addends are the same (2 + 2, 3 + 3, and so on), and *combinations of 10*, facts with addends that add up to 10 (1 + 9, 2 + 8, 3 + 7, and so on). National standards call for children to achieve phase 2 (fluency) with facts within 10 and to be able to apply strategies beyond counting to solve facts within 20 by the end of first grade. The instructional strategies and practice that *Everyday Mathematics* provides allows to children reach phase 3 (automaticity) with both the doubles and combinations of 10 by the end of first grade—an expectation beyond that suggested in the national standards.

An important achievement children reach during phase 2 is the ability to figure out facts they don't know by using a "helper fact" that they do know. For developing fluency with both addition and subtraction, the most important helper facts are doubles and combinations of 10. Beginning midway through first grade, *Everyday Mathematics* uses carefully designed Quick Looks activities with double ten frames to encourage children to develop fact strategies based on helper facts, such as *near-doubles* and *making 10*. For example, to elicit the near-doubles strategy from the class, a teacher might show two Quick Look cards in sequence, the first with eight dots in two blocks of four, and the second with nine dots in a block of four and a block of four plus one. Children describe remembering the first image of the double 4 + 4 and adding 1 more to get 9 for the second, "near to a double" image.

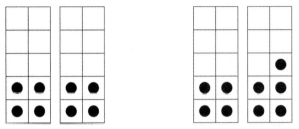

These two Quick Look cards are used to elicit the near-doubles strategy from students. Children can think of 4 + 5 as 4 + 4 + 1 and use the easier fact 4 + 4 = 8 to help them determine the total number of dots. © McGraw-Hill Education.

Everyday Mathematics uses a similar process to help children develop another critical strategy called making 10. When using making 10, children solve an unknown fact by breaking apart one addend to create a combination of 10 and then adding what remains to 10.

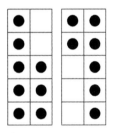

Using the making-10 strategy children imagine breaking apart the ten frame on the right and mentally moving two dots to fill the empty spaces in the one on the left, creating a 10. This way, it is easy for children to see that the sum of the two ten frames is 8 + 7 = 10 + 5 = 15. © McGraw-Hill Education.

A major priority for children progressing through first and second grade in *Everyday Mathematics* is learning to connect these strategies to the number sentences that fit the Quick Looks visualizations. For example, for the Quick Look cards illustrating the making-10 strategy, the class might record the following set of number sentences:

$8 + 7 = ?$
$8 + 2 + 5 = ?$
$10 + 5 = 15,$ so $8 + 7 = 15$

After considerable practice decomposing the dots on the Quick Look cards, children begin to apply these strategies without the aid of the double ten frames. Because most challenging addition facts can be solved using either near doubles or making 10, developing these strategies is a major step for children toward mastering their basic facts.

To help develop fluency with subtraction, *Everyday Mathematics* introduces children to *fact families*, or sets of related addition and subtraction facts, such as 4 + 7 = 11, 7 + 4 = 11, 11 − 7 = 4, and 11 − 4 = 7. Knowing how to work with fact families is particularly important for children, since a common and effective strategy for solving a subtraction fact is to think of the related addition fact, which is typically easier for children to remember. Children start working with sets of related facts like these in first grade, when they are introduced to Fact Triangles, the *Everyday Mathematics* version of flash cards. Because Fact Triangles integrate addition and subtraction, they tend to work better for children. Children also learn to use them as a self-assessment tool, sorting them into piles of facts they know and facts they need to practice.

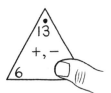

6, 7, 13 Fact Triangle

If your child is working with Fact Triangles, you can help by covering the number below the dot (the sum) with your thumb or finger and asking your child for the sum. In the example shown, you would cover the 13, and your child would practice adding 6 and 7. To help your child practice subtraction, cover either of the other two numbers. In this example, covering the 6 would encourage your child to practice thinking of either 13 − 7 = ? or 7 + ? = 13. Asking questions such as *How did you figure it out?* or *Can you think of a strategy that might help you?* will further enrich your child's practice experience.

Throughout their first few years of *Everyday Mathematics*, children have many varied opportunities to practice their addition and subtraction facts. Doing Quick Look activities, they continue working on and describing fact strategies, which they apply when playing games, and discuss repeatedly as a class. As a result, children leave second grade having achieved automaticity (solving the fact within three seconds by recall from memory or use of an automatic strategy) with their addition facts and fluency (solving the fact using an efficient and appropriate strategy) with their subtraction facts. These expectations are in line with national recommendations and define an important prerequisite for developing computational fluency with multidigit computation in later grades.

DEVELOPING FLUENCY WITH
MULTIPLICATION AND DIVISION FACTS

Everyday Mathematics introduces children to multiplication near the end of second grade, initially in the context of equal-groups number stories. For example: *Tommy has 3 packs of gum with 5 sticks of gum in each pack. How many sticks of gum does Tommy have in all?* Children typically solve such problems by drawing sketches showing the equal groups (packs of gum). They might draw 3 circles, each containing 5 dots; 3 rectangles, each split into 5 sections; or 3 sets of tally marks, with 5 tallies in each set. Children also learn to represent equal-grouping problems with arrays (rectangular arrangements of dots in rows and columns) in which each row corresponds to a group. For the problem above, children might create an array with 3 rows and 5 columns, showing the 3 groups of 5.

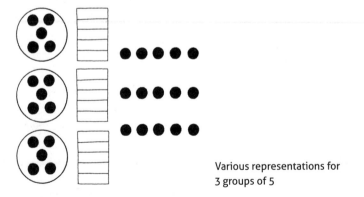

Various representations for
3 groups of 5

Once children understand the meaning of multiplication in terms of equal groups, they begin work on early multiplication facts involving 2s, 10s, and 5s. Children start making sense of these facts by relating them to their prior experiences:

- 2s facts are related to addition doubles. For example: 2 × 3 = 3 + 3 = 6 or 2 × 8 = 8 + 8 = 16
- 10s facts are related to work with money (dimes) and to skip counting by 10s (counting 10, 20, 30, 40, and so on).
- 5s facts are related to work with money (nickels) and to skip counting by 5s (counting 5, 10, 15, 20, 25, and so on).

Practice with the 2s, 10s, and 5s facts is a major focus of early third grade, since these facts will become the key "helper facts" for children's multiplication strategies. During their early exposure to multiplication, children tend to solve problems by using drawings, skip counting, or repeated addition (for example, solving 4 × 5 by adding 5 + 5 + 5 + 5). As their understanding deepens, children begin to reflect on the efficiency of these different strategies, recognizing that skip counting by familiar numbers or using repeated addition is more efficient than drawing a sketch and counting by 1s.

Once children become reasonably fluent with their 2s, 10s, and 5s facts, they are ready to begin using these facts to solve new facts. *Everyday Mathematics* encourages children to develop multiplication strategies by presenting them with pairs of linked number stories in which the information gained from solving the first story is used to solve the second one. The following is an example of such paired stories:

Ellie has 2 boxes of crayons with 8 crayons in each box. How many crayons does Ellie have in all?

Ellie's brother now gives her another box containing 8 crayons. Use what you know from the first problem to figure out how many crayons Ellie has now.

By discussing the problem as a class, children come to realize the more efficient way to solve the second problem is not by starting over from scratch, but by taking the easier fact from the first problem, 2 × 8 = 16, and simply adding on one more group of 8 and finding 16 + 8 = 24. Understanding this *adding-a-group* strategy, children apply it to solve the problem involving 3s facts by adding a group to the related 2s fact, as in this example. They might also solve 6s facts by adding a group to the related 5s fact. For example, they might solve 6 × 8 by starting with 5 × 8 = 40 and then adding on one more group of 8 to get 40 + 8 = 48, so 6 × 8 = 48.

In a similar way, children learn to work with the *subtracting-a-group* strategy by subtracting a group from a known helper fact. For example, they might use their knowledge of 10s facts to solve their 9s facts, solving 9 × 7 by starting with 10 × 7 = 70 and subtracting one group of 7 to get 70 − 7 = 63. Likewise, they solve 4s facts by subtracting a group from the related 5s fact, such as solving 4 × 6 by starting with 5 × 6 = 30 and subtracting one group of 6 to get 30 − 6 = 24.

One of the most common challenges children face working with strategies like these is deciding which number to add or subtract. For example, if you are solving 9 × 8 by starting with 10 × 8 = 80, do you subtract an 8 or a 9? Translating into the language of equal groups makes the decision easier: *I want to find out what 9 groups of 8 is. I start with 10 groups of 8, because I know 10 × 8 = 80. But I'm supposed to have 9 groups of 8, not 10 groups of 8. So I need to subtract 1 group of 8. I know that 80 − 8 = 72, so 9 × 8 = 72.* If you are helping your child solve multiplication problems, be sure to incorporate the idea of equal groups into your conversation.

Later in third grade, children begin developing additional strategies for the remaining multiplication facts. Two of the most important strategies are *doubling* and *break apart*. Using drawings to help them visualize and keep track of the steps they are taking, children tackle challenging facts by breaking them into smaller helper facts that are easier to solve. For example, children might solve 7 × 6 in either of the following ways.

Doubling: I want to solve 7 × 6. I know half of 6 is 3 and 7 × 3 = 21. So I can double 21 to find 7 × 6. 21 + 21 = 42, so 7 × 6 = 42.

Breaking apart: I can solve 7 × 6 by breaking the 7 into 5 and 2, because I know my 2s and 5s facts. 7 × 6 = 5 × 6 + 2 × 6. I know 5 × 6 = 30 and 2 × 6 = 12, so 7 × 6 = 30 + 12 = 42.

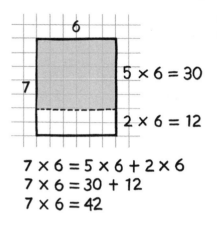

A representation of the break-apart strategy

5 × 6 = 30

2 × 6 = 12

7 × 6 = 5 × 6 + 2 × 6
7 × 6 = 30 + 12
7 × 6 = 42

Everyday Mathematics follows up on children's initial experiences with multiplication as equal groups by introducing the concept of division. Children first make sense of division by representing and solving number stories in which they equally distribute a given quantity among a given number of groups (equal sharing) or by creating equal-sized groups out of that quantity being distributed (equal grouping). As always, children begin with number stories involving familiar contexts. For example: *Ben has 24 pairs of scissors to share equally among 4 tables. How many scissors should he put on each table?* As children develop fluency with multiplication, it is natural for them to start solving division facts by thinking of the related multiplication fact. For example, they might solve 24 ÷ 4 = ? by thinking 4 × ? = 24. *Everyday Mathematics* facilitates this process of understanding by returning to Fact Triangles in third grade, this time for multiplication and division. This means that as children gain automaticity with their multiplication facts, they become fluent with the related division facts as well.

By the end of third grade, children are expected to achieve automaticity with all multiplication facts and fluency with all division facts. Just as with addition in second grade, children have many opportunities throughout grade 3 *Everyday Mathematics* to progress to phase 3 (automaticity) with their multiplication facts. They engage in meaningful practice of multiplication facts in a variety of forms.

WHERE ARE THE TIMED TESTS?

Many people are surprised to learn that *Everyday Mathematics* incorporates no timed testing of basic facts at any grade level. This may be very different from what you may remember from your own school days, but the change has been made with good reason. Research in mathematics education, psychology, and even neuroscience has documented a number of negative effects of timed testing of basic facts, in particular the onset of "math anxiety" at a very young age. Instead of timed tests, *Everyday Mathematics* incorporates a large variety of alternative assessments, which engage children in meaningful practice while allowing teachers to accurately assess all four components of fluency (flexibility, efficiency, accuracy, and appropriate strategy use). Teachers are thus able to monitor children's progress with their basic facts and complete comprehensive facts assessments without causing unnecessary and counterproductive stress. As a result, children emerge from their first few years of school with confidence, fact fluency, and a deep understanding of addition, subtraction, multiplication, and division as meaningful and useful operations. At that point, they are ready to apply their well-developed strategies to problems with larger numbers in fourth grade and beyond.

FURTHER READING

Baroody, A. (2006). Why children have difficulties mastering the basic number combinations and how to help them. *Teaching Children Mathematics, 13*(1), 22–31.

Boaler, J. (2012, July 3). Timed tests and the development of math anxiety. *Education Week.* Retrieved from http://www.edweek.org/ew/articles /2012/07/03/36boaler.h31.html

Kling, G. (2011). Fluency with basic addition. *Teaching Children Mathematics, 18*(2), 80–88.

Kling, G., & Bay-Williams, J. M. (2014). Assessing basic fact fluency. *Teaching Children Mathematics, 20*(8), 488–497.

———. (2015). Three steps to mastering multiplication facts. *Teaching Children Mathematics*, 21(9), 548–559.

Thornton, C. A. (1978). Emphasizing thinking strategies in basic fact instruction. *Journal for Research in Mathematics Education*, 9(3), 214–227.

National Research Council. (2001). *Adding it up: Helping children learn mathematics*. J. Kilpatrick, J. Swafford & B. Findell (Eds.). Mathematics Learning Study Committee, Center for Education, Division of Behavioral and Social Sciences and Education. Washington, DC: National Academy Press.

Moving On Up . . . Operations with Bigger Numbers

When most adults think of adding, subtracting, multiplying, or dividing with larger numbers, they automatically think of *algorithms*—that is, if they aren't already reaching for a calculator, a computer, or a cell phone. Algorithms, step-by-step procedures that are followed in essentially the same order every time they are used, are the primary focus of traditional math curricula. Most adults who learned mathematics in a traditional program learned a single algorithm for each operation: the U.S. traditional algorithms for addition, subtraction, multiplication, and division. If you remember "borrowing" to subtract or "carrying" to add or multiply, or the many hours spent practicing long division, you were probably taught the U.S. traditional algorithms.

The U.S. traditional algorithms for the four operations

Current research has started to detail the many challenges children face while trying to master the U.S. traditional algorithms and the many hours of valuable instructional time traditionally lost to rote drilling on these methods. Moreover, most adults who were taught algorithms in this way never understood why and how the algorithms work—and in today's world, being able to calculate without understanding is not enough. Yet many mathematics programs and teachers continue to teach only the traditional algorithms, often explaining simply, "It's how we've always done it." Maybe so. But would you accept such a rationale from your doctor in support of an outdated medical procedure when research suggests that a more effective one had been found? Is the fact that their grandparents learned to do paper-and-pencil arithmetic in a certain way reason enough to teach today's children that same arithmetic in that same way?

Instead of teaching students to mechanically follow the steps of only a single algorithm for each operation, *Everyday Mathematics* teaches children to compute accurately, efficiently, and flexibly using a variety of algorithms. The overall approach extends basic fact strategies to multidigit numbers and uses children's understanding of place value and the properties of the operations to develop many meaningful methods for multidigit computation. These methods are then formalized into a variety of carefully designed algorithms. The algorithms chosen by the authors for *Everyday Mathematics* are easier for children to make sense of, help develop their number sense and understanding of place value, and lead to better accuracy than the U.S. traditional algorithms. By teaching a variety of algorithms for each operation, with new ones introduced over time as they become developmentally appropriate, *Everyday Mathematics* gives children the power to choose which method works best for a given problem. To round out children's work with algorithms, *Everyday Mathematics* introduces the four U.S. traditional algorithms, although experience shows that many children prefer not to use them, preferring the ease of use and accuracy of the other approaches they already learned.

The way *Everyday Mathematics* teaches computation is often new to parents, but after you understand the various approaches, you may find that they are actually closer to how you naturally think about mathematics. We often hear parents say they finally understand multiplication or division after working on homework with their child. This section explains how children first make sense of and attempt to solve computation problems. This section also describes in detail one example of an algorithm that *Everyday Mathematics* presents for each operation.

MULTIDIGIT ADDITION

In *Everyday Mathematics* classrooms, children's first experiences adding multidigit numbers involve adding a 1-digit or 2-digit number and a multiple of 10 (for example: 12 + 20 = ?). Children begin adding such numbers in first and second grades, using tools such as base-10 blocks, number grids, and number lines. Children progress from using manipulatives such as base-10 blocks to solving such problems mentally, for example, by counting on by 10s (12, 22, 32).

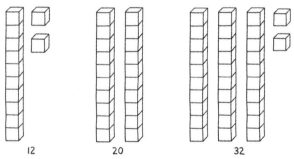

Base-10 blocks can be used to represent and combine numbers being added in multidigit addition problems.

1	2	3	4	5	6	7	8	9	10
11	12	13	14	15	16	17	18	19	20
21	22	23	24	25	26	27	28	29	30
31	32	33	34	35	36	37	38	39	40

With each 10 added, children move down another row on the number grid.

In second grade, children encounter 2-digit addition problems, such as 34 + 23 = ?, and a new tool, an *open number line*, is introduced. An open number line helps children keep track of the steps they take while adding two numbers. An open number line is similar to the traditional number line you may remember, but without so many tick marks. The only marks needed are the ones directly used for a given strategy when solving the problem, and because they are used to keep track of strategies, not for measuring, the placement of the tick marks need not be exact.

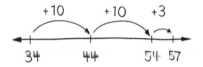

$34 + 10 + 10 + 3 = 57$, so $34 + 23 = 57$

Using an open number line for 34 + 23, children might record their steps as shown here.

Because open number lines are easy to draw, they are versatile tools for working with larger numbers. They also support mental computation. By being relieved of the need to record irrelevant details, children can focus on calculating in ways that make sense to them, which helps them develop computational fluency. In particular, because children can apply a variety of different strategies for solving a single problem, open number lines promote flexibility. In sum, open number lines promote efficient and accurate computation while providing the flexibility children need to survey the strategies they have available and to choose the ones that make sense for the problem at hand.

Later in second grade, children create their own procedures for multidigit addition, based on understanding the meaning and properties of addition and reasoning through a problem. Below are three common strategies that children might use for solving 39 + 27:

• Adding up in chunks from one number; for example, solving 39 + 27 by starting from 39, adding 20 more to get to 59, and then adding 7 more to get to 66. Children often use an open number line to keep track of the steps.

- Adjusting the numbers to make the addition easier and then compensating as needed. To solve 39 + 27, children might begin by changing 39 to 40 and adding 40 + 27 = 67. They then compensate by taking away the 1 they added to 40 to start with, leading to the final answer of 66.
- Breaking apart the numbers by place value (into tens, ones, and so on) and then combining tens with tens, and ones with ones. For 39 + 27, they would think 30 + 20 = 50, 9 + 7 = 16, and 50 + 16 = 66.

Toward the end of second grade, children formalize the breaking-apart strategy into an algorithm called *partial-sums addition*. With partial sums, children can work in whatever order makes the most sense to them. For example, whether they add the hundreds first, then the tens, then the ones, or instead add the ones first, then the tens, then the hundreds, they still end up with the same answer, because the same partial sums result either way. However, children typically begin with the largest quantities first—because that's how we read numbers and because the largest quantities are the most important anyway. Many children, given sufficient practice, are able to apply partial sums mentally, making it a much more valuable tool than U.S. traditional addition, which is generally too difficult for most people to do mentally.

Partial Sums: 137 + 65 = ?

137 → 100 + 30 + 7 65 → 60 + 5	First write addends in expanded form.
100 + 0 = 100 30 + 60 = 90 7 + 5 = 12	Then find the partial sums. Add together hundreds with hundreds, tens with tens, and ones with ones.
100 + 90 + 12 = 202	Then add the partial sums to find the total.
So 137 + 65 = 202	

The problem 137 + 65 = ? solved using partial-sums addition.

MULTIDIGIT SUBTRACTION
Children begin exploring multidigit subtraction in *Everyday Mathematics* in much the same way they started with multidigit addi-

tion, using tools such as base-10 blocks, number grids, and number lines. In first grade, children begin subtracting 10 or multiples of 10; in second grade, they are introduced to the *counting-up* method for subtraction. This approach encourages children to view subtraction as finding the distance between two numbers. To apply this approach, children begin at the smaller number and count up by 1s, 10s, or 100s to the larger number, thus "decomposing" the difference into manageable chunks. They then find the total amount that they had to count up, which is the total distance between the two numbers. Open number lines are useful for keeping track of the jumps and the total distance.

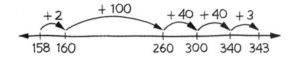

$2 + 100 + 40 + 40 + 3 = 185$, so $343 - 158 = 185$

Third graders might solve the problem 343 − 158 = ? using the counting-up strategy as shown here: 2 + 100 + 40 + 40 + 3 = 185, so 343 − 158 = 185.

Counting up is a powerful subtraction strategy for children because it relies on addition, which children tend to work with more accurately than subtraction. When children have opportunities to share their strategies and representations in class discussions, they discover more efficient ways of using counting up, such as by taking larger jumps mentally or on their number lines. Counting up can be a very efficient and accurate subtraction method for many children.

At the end of second grade, children are introduced to another method, *expand-and-trade* subtraction. Using expand-and-trade subtraction, children can complete trades (trading 1 ten for 10 ones or 1 hundred for 10 tens) in any order, and once the trades are completed, children can also perform the subtraction of corresponding hundreds, tens, and ones in any order. This way, they are able to concentrate on making trades before thinking about subtracting, which is easier than switching back and forth between trading and subtraction, as required by the U.S. traditional algorithm. Expand-and-trade also allows children to work with the larger quantities first (that is, hundreds before ones), which research shows children strongly prefer over working from the smaller quantities to the larger, which is unavoidable in the U.S. traditional algorithm.

Expand and Trade: 246 − 187 = ?

$246 \rightarrow 200 + 40 + 6$
$-187 \rightarrow 100 + 80 + 7$

First write the numbers in expanded form.

$100 \quad 140$
$246 \rightarrow \cancel{200} + \cancel{40} + 6$
$-187 \rightarrow 100 + 80 + 7$

Check each set of corresponding numbers to see if you need to make a trade. For the hundreds, 200 > 100, so you don't need to make a trade. But for the tens, 40 < 80, so you need to make a trade. Trade 1 hundred for 10 tens. For the ones, 6 < 7, so you need to make a trade. Trade 1 ten for 10 ones.

130
$100 \quad \cancel{140} \quad 16$
$246 \rightarrow \cancel{200} + \cancel{40} + \cancel{6}$
$-187 \rightarrow 100 + 80 + 7$

$0 + 50 + 9 = 59$

Finally, subtract the hundreds, the tens, and the ones. The answer is 59.

Using expand-and-trade subtraction to solve 246 − 187 = ?

MULTIDIGIT MULTIPLICATION

Midway through third grade, children develop the break-apart strategy for multiplication, using rectangular area models to help them keep track of their thinking as they solve challenging multiplication facts. (For an example of the break-apart strategy, see "There's Nothing 'Basic' about Mastering Basic Facts.") At the end of third grade, children extend the strategy to larger numbers, breaking apart the larger factor into easier numbers to multiply, and illustrating what they are doing by breaking apart the area of a corresponding rectangle.

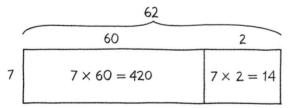

Using break-apart multiplication to solve 7 × 62 = ?

Based on their experience with the break-apart strategy, fourth graders continue their exploration of multiplication through the partial-products algorithm, in which factors are broken apart by place value (into hundreds, tens, ones, and so on). Each part of one factor is then multiplied by each part of the other factor, and, finally, all the resulting partial products are added together.

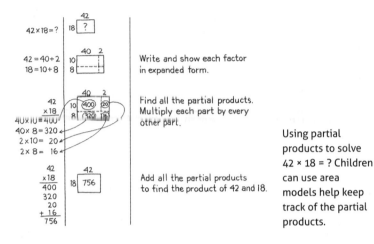

Using partial products to solve 42 × 18 = ? Children can use area models help keep track of the partial products.

Partial-products multiplication offers several advantages over other multiplication algorithms, one being that you can multiply in any order to produce the partial products. If children prefer starting with the larger quantities, as most do, they have the flexibility to do so. And because numbers are written in their entirety when working with partial products, recording 8 × 40 = 320 below the line, for example, it is easier for children to keep track of the values of each partial product. In U.S. traditional multiplication, by contrast, one must mix mental and written computation in an exact sequence, finding the product 8 × 4 = 32, writing the 2 in below the line (adding a carried digit if necessary first), and the 3 above the numbers in the next column over—a significantly more complicated procedure. Along these lines, using partial products allows children to complete all the multiplication steps required by the problem before adding together the partial products. This allows children to concentrate on one operation at a time, as opposed to alternating between multiplying and adding in the U.S. traditional algorithm. And finally, partial products is similar in structure to some techniques used in algebra, such as multiplying polynomials, and thus using this algorithm helps prepare them for later formal study of algebra.

MULTIDIGIT DIVISION

Children first encounter multidigit division in fourth grade. They begin by working with basic facts to solve related division problems that *Everyday Mathematics* calls *extended facts*. For example, children can use their knowledge that $56 \div 8 = 7$ to help them solve the problem $560 \div 8 = ?$ Children use extended facts and rectangular area models to help make sense of the partial-quotients algorithm, which is the first division algorithm taught in *Everyday Mathematics*. With partial quotients, children use easy multiples (such as 10s and 100s) to answer the question: *How many times can the smaller number go into the larger number?* Children chip away at the total quantity to be divided using the easy multiples (or "partial quotients") until they have completed the problem. They keep track of their steps using either symbolic notation or area models.

In the example $156 \div 12 = ?$ (see illustrated example), a first partial quotient of 120 (12×10) might be used to greatly reduce the total quantity being worked with. It is necessary to note, however, that the steps in partial quotients are flexible. So, for example, children who know that $12 \times 12 = 144$ might start with that partial quotient instead. Regardless of the partial quotients chosen, the final answer, once all the partial quotients are added together, will be the same.

Partial quotients offers a number of advantages over other division algorithms, including "long division," as the U.S. traditional algorithm is known. These advantages include:

- Children write out the full numbers, making place value more transparent. In the example given above, they might record "120" as the product of 12×10, not simply "12" as would be the case in the U.S. traditional algorithm. Working like this with the full number means children don't have to worry about which number to "bring down" after they subtract.

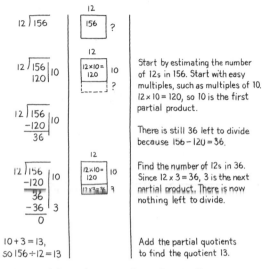

Start by estimating the number of 12s in 156. Start with easy multiples, such as multiples of 10. 12 × 10 = 120, so 10 is the first partial product.

There is still 36 left to divide because 156 − 120 = 36.

Find the number of 12s in 36. Since 12 × 3 = 36, 3 is the next partial product. There is now nothing left to divide.

10 + 3 = 13, so 156 ÷ 12 = 13

Add the partial quotients to find the quotient 13.

Using partial quotients to solve 156 ÷ 12 = ?

- Children record their partial quotients on the side as they go, so there is no need to worry where to place the digits representing their partial quotients above the division symbol.
- Children use multiples they are familiar with, which helps make the process more efficient and accurate. And as children's knowledge of multiplication grows, they can progress toward using fewer steps.

A great deal of time in elementary school is spent adding, subtracting, multiplying, and dividing whole numbers. In *Everyday Mathematics*, this work begins with number stories and progresses to strategies for basic facts. Children build on this foundation to solve problems involving larger numbers, using place value and properties of operations to compute with multidigit numbers in a variety of ways, including with the standard algorithms that have been traditional in U.S. schools for so many years. Developing true computational fluency is a high priority for *Everyday Mathematics*. From kindergarten through sixth grade, the program helps children attain that goal through the strategies and algorithms that research shows are most flexible, accurate, and efficient.

FURTHER READING

Fuson, K. C., & Beckmann. S. (2012-2013). Standard algorithms in the Common Core State Standards. *NCSM Journal of Mathematics Education Leadership, 14*(2), 14-30.

Kamii, C., & Dominick, A. (1998). The harmful effects of algorithms in grades 1-4. In L. J. Morrow & M. J. Kenney (Eds.), *The teaching and learning of algorithms in school mathematics* (1998 Yearbook). Reston, VA: National Council of Teachers of Mathematics.

National Research Council. (2001). *Adding it up: Helping children learn mathematics.* J. Kilpatrick, J. Swafford & B. Findell (Eds.). Mathematics Learning Study Committee, Center for Education, Division of Behavioral and Social Sciences and Education. Washington, DC: National Academy Press.

Verschaffel, L., Greer, B., & De Courte, E. (2007). Whole number concepts and operations. In F. K. Lester Jr. (Ed.), *Second handbook of research on mathematics teaching and learning.* Reston, VA: National Council of Teachers of Mathematics.

Algebra in Kindergarten?

Back in the 1980s, when the authors of *Everyday Mathematics* were first conceiving the program, many educators thought it would be a mistake to introduce algebra in the primary grades. Along with many parents, these educators believed children weren't ready to learn algebra at such a young age and that algebra was simply beyond the reach of most young children. The *Everyday Mathematics* authors disagreed. The research they conducted led them to believe that work with algebra should actually begin in kindergarten, allowing children to build the foundation needed for success in future formal algebra courses in middle or high school. The success young children had making sense of algebraic ideas as the authors developed the first edition of *Everyday Mathematics* confirmed their belief, and the early focus on algebra has been continued through all subsequent editions.

Today, most educators agree that even young children should be exposed to algebra, so including algebra in elementary mathematics programs has now come to be expected. Nevertheless, algebra is generally underrepresented in traditional math programs and is often included only as an afterthought or by merely labeling scattered problems as "algebraic thinking." As with the other content areas discussed in this section, algebra is approached thoughtfully and systematically in *Everyday Mathematics*, making the strand different than it is in most traditional math programs.

Perhaps you remember algebra as a mess of confusing word problems and complicated equations filled with letters and symbols. You may remember thinking, "Why am I taking this class? When will I ever use this again?" But algebra is much more than just symbols. Algebra is a part of mathematics that uses, among other things, patterns, functions, variables, equations, and modeling. The aim of this section is not to provide the full algebra trajectory in *Everyday Mathematics*. Rather, it is to share some of the unique ways *Everyday Mathematics* approaches algebra.

MODELING WITH SITUATION DIAGRAMS

In mathematics, *represent* means to show, symbolize, or stand for something. For example, a number can be represented by base-10 blocks, spoken words, or written numerals. *Modeling* is a type of representing in which what is being represented is a situation outside of mathematics, often a situation in the real world. Modeling often uses algebra—and being able to model a situation using an algebraic equation is, for most students, harder than carrying out the routine manipulations required to solve that equation. *Everyday Mathematics* has an early focus on modeling that will help students when they study algebra formally in higher grades.

Early in kindergarten *Everyday Mathematics*, children learn to model number stories and solve them concretely—by acting them out, using counters, or drawing. Consider this number story: *Madison has 2 white cars and 1 black car. How many cars does Madison have in all?* Children might model this number story by acting it out using toy cars, using 2 counters of one color and 1 counter of another color to find a total of 3 counters, or by drawing a picture.

Modeling the number story

As children become more experienced with number stories, they become more efficient in modeling them. By first grade, children use situation diagrams to model and solve different types of number stories. (See "Number Stories: The Foundation for Success with Computation" for information on number story types.) And by using these diagrams, they take an important step toward using the more abstract mathematical language of numbers and symbols. Children start off working with a change diagram, which helps them organize and make sense of the information presented in a change situation (the *starting value*, amount of *change*, and *ending value*).

Change diagram modeling this number story: *Jonah had 9 baseball cards. He lost 3 of them. How many baseball cards does Jonah have left?*

In addition to change diagrams, children are introduced to two other situation diagrams in first grade. They use parts-and-total diagrams to represent situations involving distinct parts that together make up a whole quantity. And, to model comparison number stories, children use a comparison diagram.

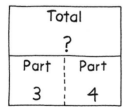

Parts-and-total diagram modeling this number story: *Christopher has 3 red pens and 4 black pens. How many pens does he have in all?*

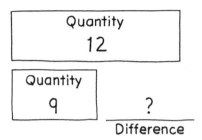

Comparison diagram modeling this number story: *Hugh has 12 pencils, and Lola has 9 pencils. How many more pencils does Hugh have than Lola?*

Situation diagrams that can be used to model multiplication and division situations are introduced in third-grade *Everyday Mathematics*. Most multiplication and division situations can be modeled using the same diagram. Later, children learn to use this same diagram to model more complicated multiplication and division situations, like those related to rates, ratios, and scaling.

Children	Books	Total Number of Books
5	6	?

Diagram modeling the following multiplication situation: *There are 5 children in a reading group. Each child must read 6 books. How many books do the children need to read in all?*

Rows	Chairs per Row	Total Number of Chairs
3	?	24

Diagram modeling the following division situation: *24 chairs need to be put into 3 equal rows. How many chairs should be in each row?*

Scale Factor	Pounds of Fertilizer	Total Pounds of Fertilizer
4	40	?

Diagram modeling the following scaling situation: *Chad needs 40 pounds of fertilizer for his lawn. Joey's lawn is 4 times as large as Chad's. How much fertilizer does Joey need for his lawn?* It is the same diagram used to model simpler multiplication and division situations.

Everyday Mathematics encourages teachers to remind children repeatedly throughout the work they do with situation diagrams that none of the diagrams are actually necessary for solving problems. Diagrams are simply devices that help organize their thinking while they are solving problems. Some children tend not to use situation diagrams or prefer to use other models, such as drawings, while others find the diagrams useful.

MODELING USING NUMBER SENTENCES AND VARIABLES

Another way to model number stories is with number sentences. A *number sentence* is like an English sentence, but one made up of numbers and symbols rather than words. Defined precisely, a number sentence must include at least two numbers or expressions separated by a relation symbol such as <, >, or =. For example, $3 + 5 = 8$ and $5 > 4$ are number sentences. An *expression* is any group of mathematical symbols that represent a number or quantity. Expressions do not contain relation symbols, so $3 + 4$ and $6 * (2 + 8)$ are both expressions but $5 = 3 + 2$ is not an expression, but rather, a number sentence and an equation. *Equations* are number sentences that contain an equal sign. $3 + 5 = 8$ is an equation but $5 > 4$ is not an equation (it's an inequality), although both are number sentences. Keeping this distinction straight is important for children learning about equality, which is a fundamental concept in mathematics. *Everyday Mathematics*, unlike traditional programs, reinforces children's learning by varying the position of the equal sign in equations. So, for example, children will regularly see both $3 + 5 = 8$ and $8 = 3 + 5$, whereas most math programs consistently use only the former.

Everyday Mathematics uses the term *number model* specifically to mean number sentences, expressions, or equations that model number stories or real-world situations. When number sentences, expressions, or equations appear on their own, isolated from any nonmathematical context, *Everyday Mathematics* does not refer to them as number models.

Children begin modeling number stories symbolically, using the plus (+), minus (−), and equal (=) symbols, in kindergarten.

Kindergartners could represent the number story from the previous section—*Madison has 2 white cars and 1 black car. How many cars does Madison have in all?*—by any of the following number models:

$$2 + 1 = 3 \qquad\qquad 1 + 2 = 3$$
$$3 = 1 + 2 \qquad\qquad 3 = 2 + 1$$

Typically, children first write number models to represent their solutions *after* they have already solved the problems, as in the example here with Madison's cars. Writing the number model is a way to summarize the story and is helpful to children as a means to check their work. Beginning in first grade, however, *Everyday Mathematics* teaches children to do something that is traditionally taught only in later grades. First graders begin to represent problem situations *before* they solve them by writing number models using *variables*.

A variable is a letter or other symbol that stands for a missing or unknown number. Using letters as variables, as in the number sentence $3y + 9 = 18$, is what most people recognize as algebra. Of course, *Everyday Mathematics* does not start children off with such complex number sentences, relying instead on the structure of the situation diagrams, which is already familiar to children and helps them write number models with variables to represent the unknown numbers. For example, consider this situation: *Taylor collects stickers. Last week she had some stickers. This week, she got 4 more stickers. Now she has a total of 7 stickers. How many stickers did Taylor have last week?* Children begin by using a change diagram to represent the problem, writing in the variable ? to represent the unknown. Then they follow the structure of the completed diagram to write a number model, now with an answer box representing the unknown addend: $\square + 4 = 7$. Children find the answer using whatever approach makes sense to them. The answer is the number that will make the number sentence true when it's put in the answer box.

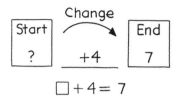

Change diagram and number model for the following situation: *Taylor collects stickers. Last week she had some stickers. This week, she got 4 more stickers. Now she has a total of 7 stickers. How many stickers did Taylor have last week?*

In *Everyday Mathematics*, first graders use blanks (__) and answer boxes (□) interchangeably as variables representing unknown numbers in number models, number sentences, and equations. In situation diagrams, ? is used to represent an unknown quantity. Beginning in second grade, the blanks and answer boxes representing variables in number models are often replaced by a question mark, such as 12 − 9 = ? or 9 + ? = 12. In third grade and beyond, unknown values are represented in number sentences and number models by letters, blanks, or question marks. For example:

5 + ___ = 12 $14 \div ? = 7$

20 = 10 + n $y = 4 * 8$

While children begin to purposefully use variables already in first grade in *Everyday Mathematics*, they are not responsible for using the term *variable* until it is formally introduced in fourth grade.

PATTERNING AND FRAMES AND ARROWS
Patterns are another important facet of algebra. When someone says "patterns," you probably think of:

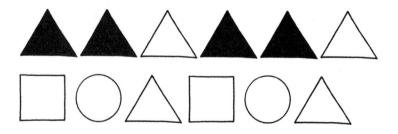

Kindergartners in *Everyday Mathematics* do explore patterns like these, but their work with patterns extends far beyond shapes and colors. They begin examining patterns on the Number Grid, for example, when they play a game called *Number Grid Cover-Up*.

−9	−8	−7	−6	−5	−4	−3	−2	−1	0
1	2	3	4	5	6	7	8	9	10
11	12	13	14	15	16	17	18	19	20
21	22	23	24	25	26	27	28	29	30
31	32	33		35	36	37	38	39	40
41	42	43	44	45	46	47	48	49	50
51	52	53	54	55	56	57	58	59	60

The game *Number Grid Cover-Up*: How can you use patterns to figure out which number is covered?

Starting in first grade, *Everyday Mathematics* introduces Frames and Arrows, a pattern activity that is unique to the program. A Frames-and-Arrows diagram shows a sequence of numbers that follow a given rule. Each frame contains a number. Each arrow represents the rule that determines the next number in the sequence.

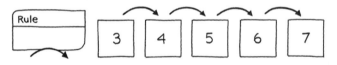

The task is for children to discover a rule for the diagram. They might describe it as "add 1," but it could also be "+ 1" or "count by 1s." Or, children might use the rule to find the missing numbers in the diagram, such as the problem below.

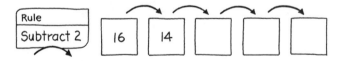

Children solve by subtracting 2 from each number in order, writing the numbers 12, 10, and 8 in the empty frames.

In a third type of Frames-and-Arrows diagram, children look at the number sequence and use it to find the arrow rule.

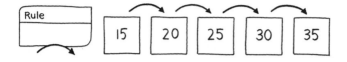

Children solve by observing the change between consecutive numbers and identifying the rule as "add 5," "+ 5," or "count by 5s" in the rule box.

Starting in third grade, Frames-and-Arrows can have more than one rule. In the following problem, the black arrow means + 1 and the gray arrow means + 2.

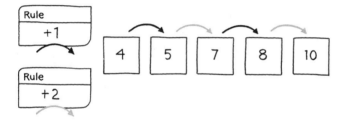

Most children are quick to understand Frames-and-Arrows diagrams, finding in them a flexible way to deal with patterns involving consistent changes, which will be important to future work with formal algebra. Moreover, Frames-and-Arrows activities help children explore the relationship between counting and addition or subtraction and develop strategies for finding unknown numbers in number sentences. For example, consider this problem:

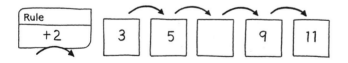

Finding the missing number in this problem is like solving $5 + 2 = \square$ or $\square + 2 = 9$.

By learning to solve Frames-and-Arrows problems, children take an important step toward developing deeper algebraic understanding. Parents of children in grades 1 through 4 can reinforce these ideas whenever Frames and Arrows comes up by taking turns with their children making up and solving Frames-and-Arrows problems of all varieties.

FUNCTION MACHINES AND "WHAT'S MY RULE?"

Another aspect of algebra involves work with functions, which can be thought of as rules that transform input values into corresponding output values in consistent ways. *Everyday Mathematics* introduces children to functions in kindergarten, beginning with a game called *What's My Rule? Fishing.* The teacher "catches" children by calling those with certain characteristics (wearing red shirts, having slip-on shoes, and so on) to the front of the class, asking, "What sort of fish am I fishing for? What is my fishing rule?" Children use their observations and reasoning to deduce what the rule must be. Using rules to create and describe categories in this way is an important mathematical skill.

Later in the year, kindergartners are introduced to an activity called "What's My Rule?" which *Everyday Mathematics* uses in all grades. "What's My Rule?" problems are based on tables of input and output numbers that are related by a rule or function. Children use the idea of a "function machine," which is a virtual, rather than a real, machine, to understand how the input and output numbers in "What's My Rule?" tables are produced. Function machines are "programmed" to take in numbers (input), to apply a rule to the input numbers, and then to send out the new numbers (output). In some "What's My Rule?" problems, children are given input numbers and a rule and are asked to find "what comes out." In other instances, children know the rule and the output number, but the input is unknown and children must find "what goes in." And in some problems, children must find the rule, based on what they notice about the given inputs and outputs. (In general, many rules can fit any given set of input and output numbers, but fortunately there is usually a "simple" rule such as "add 1" or "double" that works. If there's ever any doubt about whether a proposed

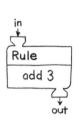

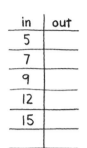

in	out
5	
7	
9	
12	
15	

What comes out?

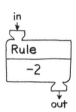

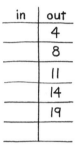

in	out
	4
	8
	11
	14
	19

What goes in?

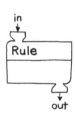

in	out
1	2
2	4
3	6
4	8
5	10

What is the rule?

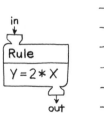

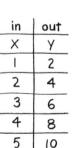

in	out
X	Y
1	2
2	4
3	6
4	8
5	10

A "What's my rule?" diagram using algebraic notation

rule is correct, it can always be checked against the given input and output numbers: If the rule produces the proper outputs, then it is correct.)

Much of the "What's My Rule?" work in early grades provides practice with basic facts. And, as with Frames and Arrows, "What's My Rule?" activities provide opportunities for children to learn to find unknown numbers in number sentences. For example, in the "What goes in?" figure, finding the first input is equivalent to solving $\Box - 2 = 4$.

In fifth grade, children begin working with rules that are written using variables. For example, in the "What is the rule?" figure, the rule is $* 2$, which means that you double the "in" number to get the "out" number. This rule can be written as $y = 2 * x$ using algebraic notation.

Rules can also be graphed on a coordinate grid. The numbers in the table become ordered pairs, where the "in" numbers are the x-coordinates and the "out" numbers are the y-coordinates. Graphing ordered pairs of corresponding values helps children visualize relationships of various sorts and extend patterns beyond the listed values. This work helps prepare students to continue exploring functions in formal algebra classes.

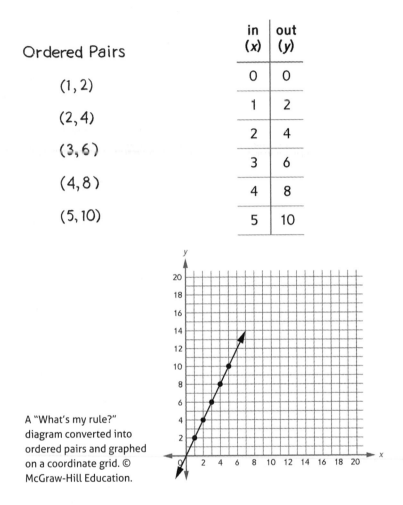

Ordered Pairs

(1, 2)

(2, 4)

(3, 6)

(4, 8)

(5, 10)

in (x)	out (y)
0	0
1	2
2	4
3	6
4	8
5	10

A "What's my rule?" diagram converted into ordered pairs and graphed on a coordinate grid. © McGraw-Hill Education.

Throughout *Everyday Mathematics*, activities involving algebra offer many opportunities for children to develop their algebraic thinking. Starting in kindergarten, these activities lay the foundations for the more formal study of algebra in middle and high school.

FURTHER READING

Usiskin, Z. (1997). Doing algebra in grades K–4. *Teaching Children Mathematics*, 3(6), 346–356.

The Trouble with Fractions

It has long been apparent that many children struggle when it comes to fractions. In fact, even many adults find fractions confusing. A recent study, for example, found that not one person in a sample of two dozen U.S. adults was able to create a word problem that correctly fit the fraction division problem $1\frac{3}{4} \div \frac{1}{2}$. Partly because of such documented difficulties with fractions and partly because facility with fractions is so important for success with algebra, a great deal of recent research has aimed at understanding how fraction concepts develop in elementary school. This research shows that major problems include incomplete understanding of what fractions are, inappropriate use of whole-number reasoning to fraction situations, and insufficient understanding of the reasoning behind common procedures for fraction computation. It should not be surprising, for example, that a person whose understanding of fraction division is limited to "invert and multiply" cannot correctly interpret $1\frac{3}{4} \div \frac{1}{2}$. Nor should it come as a surprise that a person who incorrectly thinks $\frac{3}{8}$ is bigger than $\frac{3}{4}$ because 8 is bigger than 4 would struggle with fractions.

The authors of *Everyday Mathematics* have studied this research to improve the treatment of fractions in every new edition of the program. The result is that work with fractions in *Everyday Mathematics* is often very different from work with fractions in other math programs. This section sketches how *Everyday Mathematics* handles the development of fraction understanding and fraction number sense.

EARLY FRACTION WORK

The differences between how *Everyday Mathematics* and other programs treat fractions begin with the early work designed to help children understand of the meaning of fractions. Most math programs introduce fractions by having children identify and color parts of shapes that have been pre-divided into equal parts. For example, given a circle divided into four equal sections, children are asked to color in one section and name the corresponding fraction ($\frac{1}{4}$). Traditional programs then move directly to describing fractions using formal standard fraction notation ($\frac{1}{2}$, $\frac{3}{4}$, $\frac{5}{8}$, and so on), often as early as first grade.

Everyday Mathematics approaches initial fraction work from a different perspective, one supported by recent research and aimed at building the conceptual basis children need to truly understand fractions. Rather than working with pre-divided shapes, first graders begin by devising their own strategies to divide, or *partition*, shapes and other objects into equal parts. By focusing on the idea of equal shares—often in contexts in which they are naturally motivated to divide things equally, as in when they share pancakes or muffins with their friends—children develop a conceptual understanding of what fractions really are and why fractions and division are connected. Children also partition paper circles and rectangles, dividing them into two (and later three, four, and more) equal parts called *shares*, and they spend ample time discussing how they know the shares are equal.

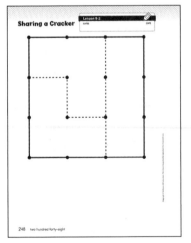

© McGraw-Hill Education.

The equal shares in these early lessons frequently have the same shape, because that makes it easier for young children to verify that the shares are equal. However, it is important to understand that it is not necessary for shares to be the same shape to be equal. *Everyday Mathematics* moves on to this important idea in second grade, as children learn to prove, for example, that dividing a cracker does, in fact, result in equal shares.

Unlike math programs that

introduce standard notation for fractions as early as first grade, *Everyday Mathematics* delays standard notation ($\frac{1}{2}$, $\frac{1}{3}$, $\frac{4}{4}$, and so on) until the middle of third grade. The reason for this is simple. Research shows that young children tend to think of standard fraction notation as two separate whole numbers, causing them to misapply whole-number reasoning to fractions. Children often believe, for example, that because 4 is greater than 2, $\frac{1}{4}$ is greater than $\frac{1}{2}$. Or they might "add across" when adding fractions, mistakenly thinking that $\frac{1}{2} + \frac{1}{2} = \frac{2}{4}$ because $1 + 1 = 2$ and $2 + 2 = 4$.

Misconceptions like these can be avoided if instruction begins by helping children understand fractions as *single numbers* rather than as combinations of two whole numbers. *Everyday Mathematics* cultivates this understanding by encouraging children working with equal shares to use and discuss alternative *written* names for the equal partitions and for the recomposed whole. Children describe the partitions and wholes in a variety of ways, including *1 out of 2 parts*, *1 quarter*, *1 half*, *2 halves*, and *4 fourths*. These terms help children distinguish between different kinds of partitions, and also emphasize the relationship between equal shares and the whole, reinforcing the idea that putting together *2 halves* (or *4 fourths*) results in *1 whole*. This way of talking about and writing fractions is used throughout first and second grades and in the beginning of third grade.

In the middle of third grade, *Everyday Mathematics* introduces standard fraction notation by connecting it with familiar pictorial and physical representations of fractions that children have worked with since first grade. For example, in the fraction $\frac{5}{8}$, 5 is the numerator and 8 is the denominator. Children learn that the *numerator* represents the number of parts being considered, while the *denominator* represents the number of equal parts needed to make 1 whole, which in turn indicates the size of the parts (since, for example, a bigger denominator means smaller parts). Thus the fraction $\frac{5}{8}$ represents 5 parts, each of which is one-eighth of the whole in size. Users of *Everyday Mathematics* report that providing children with a more solid understanding of fractions as a single number, and introducing standard fraction notation later in the curriculum, does in fact help prevent misconceptions.

FRACTION REPRESENTATIONS

In most mathematics programs, children encounter a wide variety of fraction representations that are not used consistently from grade to grade. Educational researchers who study fractions recommend using fewer and more consistent representations of fractions when children are still developing their initial fraction ideas and as they are developing strategies and procedures for fraction operations. Working with a smaller number of carefully selected models gives children time to gain confidence and deepen their understanding of fractions and their relative sizes. Once children become comfortable with at least one of these models, they find it easy to visualize in their heads what the various fractions look like. Rather than thinking that $\frac{1}{4}$ is more than $\frac{1}{2}$ because 4 is more than 2, to return to our previous example, they can call upon their mental models of fractions and can "see" that $\frac{1}{2}$ is more than $\frac{1}{4}$. So *Everyday Mathematics* uses a small, consistent, standardized set of representations across the treatment of fractions in grades 3–6, including fraction circle pieces, fraction strips, and fraction number lines, as described below.

FRACTION CIRCLE PIECES

Beyond the finding that using fewer models leads to better fraction understanding, research also suggests that some models are better than others in supporting concept and skill development. *Fraction circle pieces* (also called *fraction circles*) are a good example. They have been shown to provide the most effective model for building the mental images children need to understand order, equivalence, and fraction operations. Fraction circles are sets of circles divided into equal-sized pieces. Each size represents a different *unit fraction* (a fraction whose numerator is 1, such as $\frac{1}{2}, \frac{1}{3}, \frac{1}{4}$, and so on), and all of the same-sized pieces are the same color. Beginning in third grade, *Everyday Mathematics* uses fraction circles to help children develop an understanding of the "whole," which they need to understand unit fractions as equal parts of that whole.

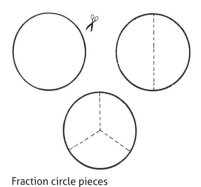

Fraction circle pieces

Many children initially see the circle as the only fraction circle piece that can be the "whole." To avoid this common misconception, *Everyday Mathematics* presents children with different shapes to represent the whole. They learn that the name of a fractional part is linked with the size of the whole it belongs to. That is, they learn the valuable lesson that the same physical object can be both one-fourth of a circle and also one-half of a semicircle. The size of the whole matters: half a mile is not the same as half an inch.

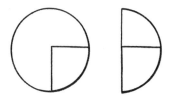

The name of the small piece depends on what has been identified as the whole. If the circle is the whole, the name for the small piece would be 1-fourth. If the semicircle is the whole, the name would be 1-half.

By using a variety of different pieces to represent the whole, children learn flexibility with fractions. This flexibility helps them to

- understand how fractions communicate a relationship between parts and a specific whole;
- recognize a fraction as a *single number* that expresses that relationship rather than two distinct, unrelated numbers;
- solve real-world problems in which the size of the whole varies; and
- understand and perform fraction operations (addition, subtraction, multiplication, and division) in later grades.

FRACTION STRIPS

Third graders in *Everyday Mathematics* are also introduced to fraction strips, which are rectangular strips of paper folded and partitioned to represent fractions. Because fraction strips are made by folding paper, children are able to connect them to their prior experiences with partitioning in first and second grade. They learn that when they fold a strip to form b equal parts, each equal part represents a unit fraction, or $\frac{1}{b}$. So when they fold a strip into 6 equal parts, each part is one-sixth of the whole strip, which is written as the unit fraction $\frac{1}{6}$. Working with fraction strips helps children recognize that non-unit fractions are a count (or sum) of unit fractions. For example, dividing a strip into 6 equal parts shows children that $\frac{1}{6}$ is a single one-sixth part, while $\frac{5}{6}$ is 5 one-sixth parts. This idea is foundational for understanding fractions as single numbers.

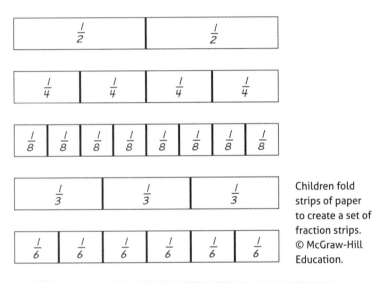

Children fold strips of paper to create a set of fraction strips. © McGraw-Hill Education.

FRACTION NUMBER LINES

Because fraction strips are rectangular in shape, working with them makes it easier for children to bridge between their early use of shapes like fraction circles and a familiar model *Everyday Mathematics* now brings back into the picture: the number line. The connections between the two models, and consequently the important concepts that unite them, become tangible for children when they see how they can align their fraction strips with the markings on a number line.

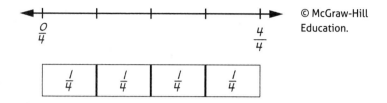

Putting fractions on a number line develops children's understanding of fractions as numbers because number lines bring whole numbers and fractions together in the same model. Being able to locate fractions and whole numbers on the same number line reinforces children's ability to see fractions as single numbers: Rather than being two numbers, 3 and 4, $\frac{3}{4}$ is itself a number between 0 and 1 because it's located to the right of 0 and the left of 1 on the number line. In third grade, children begin to understand that fractions can also represent distances on the number line, just as fractions on a ruler represent lengths. Children understand the whole on a number line as the distance between 0 and 1. As they move along the number line, they connect fractions with distances traveled from 0.

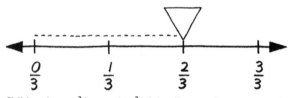

Children locate $\frac{2}{3}$ by moving $\frac{2}{3}$ of the distance from 0 to 1, or by partitioning the distance between the points 0 and 1 into three equal parts and then counting two of those parts.

Children also learn to visualize and represent fractions equal to or greater than 1 by working with the number line. When 1 is represented as the fraction $\frac{3}{3}$ on the number line, children simply continue counting by thirds to locate $\frac{4}{3}$, $\frac{5}{3}$, and so on. In general, a fraction $\frac{a}{b}$ can be thought of as made up of a parts, each of which is $\frac{1}{b}$ in length. For example, $\frac{3}{4}$, or three quarters, can be thought of as 3 parts, each of which is $\frac{1}{4}$, or one quarter, in length. Third graders work to understand that the fraction $\frac{a}{b}$ refers both to the interval on the number line of length $\frac{6}{5}$, with its left endpoint at 0, and to the point on the number line at the end of that length. Thus the number $\frac{6}{5}$ is both a length and a single point on the number line. These ideas are crucial for the mathematics children will learn in fourth grade and beyond.

Using a number line also helps children make sense of equivalences between fractions greater than 1 and mixed numbers, so that they see, for example, that three-halves ($\frac{3}{2}$) and one and one-half ($1\frac{1}{2}$) are equivalent. Children grasp the concept of equivalency more easily once they recognize that the interval between any two consecutive whole numbers can be divided into equal parts corresponding to the denominator of a fraction. They understand that they can travel 1 whole unit and 1 half unit from 0 to locate $1\frac{1}{2}$. Or they can travel 3 half units from 0 to locate $\frac{3}{2}$. Either way results in the same ending point, so children understand that $1\frac{1}{2}$ is equivalent to $\frac{3}{2}$.

Finally, working with number lines reinforces yet again the importance of carefully identifying the unit whole. Two fractions or decimals can only be compared if the unit whole is the same for both. Children see that while $\frac{3}{4}$ inch is greater than $\frac{1}{2}$ inch, $\frac{3}{4}$ inch is less than $\frac{1}{2}$ mile because the unit wholes are different.

DEVELOPING STRATEGIES
Beginning in third grade and continuing into sixth grade, children work with fractions in any number of ways, including finding equivalent fractions, comparing and ordering fractions, and computing with fractions (adding, subtracting, multiplying, and dividing fractions). Doing all of this entails tasks such as the following:

- finding other fractions that are equivalent (equal) to $\frac{2}{3}$, such as $\frac{4}{6}$, $\frac{6}{9}$, and so on;
- comparing two fractions to determine which is larger. For example, children might recognize that $\frac{2}{3}$ is larger than $\frac{1}{4}$ because $\frac{2}{3}$ is greater than $\frac{1}{2}$, while $\frac{1}{4}$ is less than $\frac{1}{2}$; and
- computing when at least one number is a fraction, such as $5 \times \frac{2}{3}$ or $\frac{3}{4} - \frac{1}{3}$.

Children (and adults) sometimes struggle with fractions because they were taught to follow rote processes without understanding. Perhaps you remember dividing fractions using the "invert-and-multiply" rule that says that to divide two fractions, you can invert (or flip over) the second fraction and multiply it by the first fraction. While you may remember doing this in school, did you actually know *why* you were doing it and *how* it works? Or were you just taught to follow the rule to get the correct answer? The fact that most adults find themselves unable to explain why this rule works means they were never given the opportunity to develop good fraction number sense when they were learning mathematics.

Another likely reason people struggle finding equivalent fractions, comparing and ordering fractions, and computing with fractions is because it can seem that the rules for working with fractions are very different from the rules for working with whole numbers. Think again about division. When dividing whole numbers, the quotient (or answer) is smaller than the dividend (the number being divided). When solving $100 \div 20$, for example, the quotient, 5, is smaller than the dividend, 100. When fractions are divided this is not always true. Indeed, dividing fractions often results in a quotient that is larger than the dividend. To children this seems counterintuitive, until they really apply their understanding of the meaning of division to fractions. For example, recognizing that $1\frac{1}{2} \div \frac{1}{4}$ can be thought of as the question *How many $\frac{1}{4}$ cups of sugar can I get out of $1\frac{1}{2}$ total cups of sugar?* helps make it seem reasonable that the answer is 6.

Many math programs approach advanced fraction work (finding equivalent fractions, comparing and ordering fractions, and fraction computation) by having children learn formal *algorithms*, or sets of step-by-step instructions for solving a problem. Often children don't fully understand *why* the steps exist, and memorizing an algorithm without actually understanding it can result in difficulties learning more advanced mathematics down the road. *Everyday Mathematics* aims for children to understand the work they are doing with fractions. And this is the reason for another big difference between *Everyday Mathematics* and other math programs: we delay the introduction of formal, standard procedures for advanced fraction work.

Rather than teaching children to memorize a series of steps, *Everyday Mathematics* prompts children to use visual models (fraction circles, fraction strips, and fraction number lines) and informal reasoning to invent their own strategies for solving fraction computation problems. In *Everyday Mathematics* classrooms, children solve problems in a variety of ways, share their solution strategies, and compare and discuss the approaches they used. It is only in later grades after much informal work that teachers take the next step, which is helping children generalize their strategies and introducing formal algorithms for addition, subtraction, multiplication, and division with fractions.

The root cause of the trouble with fractions for many children and adults is not fully understanding the most fundamental thing about fractions—what a fraction really *is*. Hopefully, this description of how *Everyday Mathematics* teaches fractions to your children has helped you to gain a firmer grasp of the concept, which will help you support your child with fraction work at any grade level. Having a solid sense of what fractions are and how fractions relate to each other helps children understand equivalence, comparison, and computing with fractions. Their fraction understanding helps them to better make sense of fraction problems, judge the reasonableness of their answers, and avoid the fraction pitfalls that have plagued students for so many years.

FURTHER READING

Cramer, K., Behr, M., Post, T., & Lesh, R. (2009). *Rational Number Project: Initial fraction ideas*. Dubuque, IA: Kendall/Hunt Publishing Co. Retrieved from http://www.cehd.umn.edu/ci/rationalnumberproject /rnp1-09.html. (Originally published in 1997 as *Rational Number Project: Fraction lessons for the middle grades—level 1*.)

Cramer, K., & Wyberg, T. (2009). Efficacy of different concrete models for teaching the part-whole construct for fractions. *Mathematical Thinking and Learning, 11*(4), 226–257.

Empson, S. B., & Levi, L. (2011). *Extending children's mathematics: Fractions and decimals*. Portsmouth, NH: Heinemann.

What Happened to Rulers?

Learning about measurement may seem straightforward. And in many ways it is. By the time children reach sixth grade in most mathematics programs, they know how to compare and measure length, perimeter, area, and angles. They are able to calculate mass and volume, convert between different units of measure, tell time, and work with money. *Everyday Mathematics* differs from other math programs, however, in the way it teaches children to understand exactly what a measurement is, what it means, and how it is derived. The treatment of length measurement illustrates the general approach, though the same approach carries across other measures. As with other content areas, we don't have the space here to describe the entire measurement trajectory by grade level, nor can we give a detailed account of how different topics within measurement, such as time, area, or volume, are addressed. Instead, we focus on what makes the *Everyday Mathematics* treatment of *early length measurement* unique, as a way of illustrating the program's general approach to measurement topics.

If your child is in kindergarten or first grade, you may have wondered what happened to the rulers you remember from grade school and why your child has not yet learned terms such as *inches, feet,* and *centimeters.* In most traditional mathematics programs, children begin learning about length in kindergarten in terms of these standard, conventional units, such as inches and centimeters. And this instruction is followed by much rote practice, in which children repeatedly line up one end of the object being measured to the zero mark on a ruler and read off the number at the other end.

From their work with researchers who study how children learn measurement, the authors of *Everyday Mathematics* found that teaching length measurement in the traditional way outlined above often results in children having faulty understandings of what length actually is, how units are used to measure length, and why we need standard units. As a result, many children experience difficulty understanding length as a distance along a path and they may not understand that a measurement of length is made up of individual units added together. These misunderstandings, in turn, lead to problems such as incorrectly measuring lengths of objects whose ends are not lined up with the zero mark on the ruler. Problems such as these reflect an overreliance on measurement procedures that children don't fully understand.

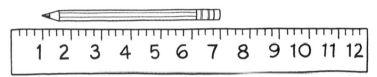

Many children incorrectly measure this pencil to be 7 inches long.

Everyday Mathematics aims to help children avoid developing these common misconceptions during its treatment of measurement in early elementary school. Rather than jumping right in with the familiar inch ruler, *Everyday Mathematics* introduces kindergartners to measurement by focusing on describing and comparing objects based on their relative lengths, as opposed to their measured lengths. They use direct comparison to order a series of objects by length. And they spend time comparing weight and capacity in similar ways.

First grade *Everyday Mathematics* also begins work with length measurement with visual comparisons of "longer" and "shorter," as children compare lengths of two objects by placing them side by side to determine which is longer. Then they learn to compare objects indirectly, using a third object as an intermediary. For example, if we know that Jose is taller than Bill and that Bill is taller than Rashid, then we know that Jose is taller than Rashid, even if Jose and Rashid never compare their heights directly. Comparison work like this, which in traditional programs can be lost to all the rote practice, helps children take an important step toward understanding *quantitative measurement*, or measurement using numbers, such as 12 centimeters, 5 yards, and so on. By learning to say that one object is shorter than another, children are essentially using the shorter object as a unit for measuring the longer object.

Near the end of kindergarten in *Everyday Mathematics* and throughout first grade, children learn to use *nonstandard units*, or informal, unconventional units that can be used repeatedly to measure lengths of objects. Examples of nonstandard units include connecting cubes, stick-on notes, toothpicks, paper clips, and pencils. By beginning with nonstandard units, as opposed to inches or centimeters pre-printed on a ruler, *Everyday Mathematics* allows children to focus not on the process of making formal measurements but on what we mean by the concept of measurement itself.

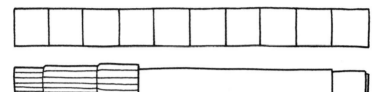

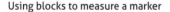
Using blocks to measure a marker

Working with nonstandard units encourages children to discover a number of important features of length measurement. For example, in first grade, children experiment with measuring the same item using a variety of different nonstandard measurement units. They might discover that a child who measured the length of a desk using large paper clips obtained a different measurement than a child who measured the same desk using small paper clips. Activities like this help children recognize the importance of using same-sized units for generating consistent measures and also the relationship between the size of the unit and the number of such units needed to fill a given length (i.e., that it takes more of the smaller paper clips than the larger paper clips to measure the desk). Children also learn that measuring involves placing units without overlapping or leaving gaps and that the entire length must be measured from one end to the other of the path or object. In other words, they learn that the space being measured must be carefully "filled up" with units. Finally, they learn how units must align to the path or object being measured. If the units don't make a straight line along the path, the measurement will be faulty. On the other hand, if all of these conditions are met, then the number of units counted is a measure of the length of that path or object.

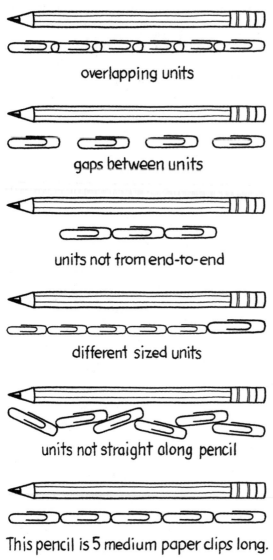

overlapping units

gaps between units

units not from end-to-end

different sized units

units not straight along pencil

This pencil is 5 medium paper clips long.

Using paper clips to measure a pencil

After children have become familiar with measuring objects and straight paths, *Everyday Mathematics* introduces the idea of crooked paths. While this idea is often missing from traditional mathematics programs, it is important to include it because it helps children understand the concept of *additivity of length*. In other words, children determine that the length of a crooked path is the same whether it is measured all at once (as with a tape measure) or in smaller pieces where the lengths are then added together.

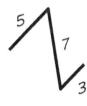

Children find the length of the path by adding together the lengths of the segments.© McGraw-Hill Education.

$$5 + 7 + 3 = 15$$

Working like this with crooked paths helps children consolidate their understanding of length measurement, which is important background for finding perimeter, or the distance around the boundary of the shape.

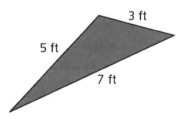

Children find the perimeter of the triangle by adding together the lengths of the sides. © McGraw-Hill Education.

$$5 \text{ ft} + 3 \text{ ft} + 7 \text{ ft} = 15 \text{ ft}$$

After all the work with nonstandard units, most first graders understand the basics of length measurement. And by the end of first grade, most children have discovered that it is easier to quickly and accurately measure length if they "collect units," that is, if they use tools composed of multiple units. First-grade work concludes with children making their own rulers with paper clips and using these rulers to measure the length of objects in paper-clip units.

Paper-clip ruler. © McGraw-Hill Education.

Children are now in a position to grasp the idea that measuring length is about answering the question: How many units does it take to fill the space along this object or path? And they understand that measuring is more than simply reading off a number from a ruler. This rich background enables them to overcome the common misconceptions described at the beginning of this section, allowing them to find the correct length of an object, no matter where the endpoint of the object being measured is aligned.

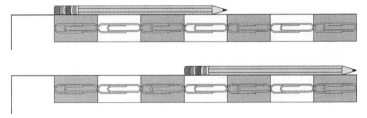

Children find length by counting the units spanned by the object or path even when neither end of the object is aligned at zero. © McGraw-Hill Education.

By second grade, children are ready to transition from measuring length using concrete, informal units to measuring length using rulers and tape measures marked in standard units such as inches and centimeters. To help children make the connection between the concrete objects they have used to measure and the unit spaces on a ruler, they line up 1-inch-square pattern blocks along an inch ruler. Then they measure objects using both blocks and rulers, demonstrating to themselves both methods produce the same measurements. Only then are they ready to begin working with length measurement the way you probably remember it, by using a ruler.

Hopefully, by understanding how *Everyday Mathematics* teaches length measurement, you have gained a better appreciation of the complexity involved in teaching this seemingly simple topic. Our approach to other measurement topics, such as area or volume, is similar. Through carefully constructed sequences of activities that help children avoid common misconceptions and build their understanding, *Everyday Mathematics* equips them with the knowledge they need to continue exploring this important area of mathematics.

FURTHER READING

Battista, M. T. (2006). Understanding the development of students' thinking about length. *Teaching Children Mathematics, 13*(3), 140–146.

Levine, S. C., Kwon, M., Huttenlocher, J., Ratliff, K. R., & Dietz, K. (2009). Children's understanding of ruler measurement and units of measure: A training study. In N. A. Taatgen & H. van Rijn (Eds.), *Proceedings of the 31st annual conference of the Cognitive Science Society* (pp. 2391–2395). Austin, TX: Cognitive Science Society.

Smith, J., Sisman, G. T., Dietiker, L., et al. (2008). *Framing the analysis of written measurement curricula*. Paper prepared for the Annual Meeting of the American Education Research Association, New York.

SECTION 3
What to Expect from
Everyday Mathematics

Now that you know a little more about what *Everyday Mathematics* is and why your district selected it as the best math curriculum for your child, you may be interested in learning what you as a parent can expect from the program. What kinds of materials will your child be bringing home? What will your child be learning from *Everyday Mathematics*?

This section details what you can expect from *Everyday Mathematics*, with the stipulation that classrooms will vary in their implementation of the program, especially as more schools move toward full digital implementation. Some classrooms will continue to have children work only on printed materials, while others will expect children to complete some or all of their work digitally, using tablets or computers. This means that what you see at home, especially whether you see work completed in print or digital formats, will vary depending on your child's school, classroom, and teacher. Be sure to talk to your child's teacher if you are unsure how and where you can see your child's work and monitor his or her progress. More information about digital *Everyday Mathematics* can be found in section 4.

Linking Math in School to Math at Home

The authors of *Everyday Mathematics* believe that the connection between home and school is very important for your child's education. For this reason, *Everyday Mathematics* provides nearly daily communication with parents in the form of *Home Links*®, the *Everyday Mathematics* version of homework. *Home Links* ask children to review and practice the content of the day's lesson. Many *Home Links* also provide additional practice, often with computation, to help children develop fluency. Since *Home Links* follow up directly on the content of the day's math lesson, they provide

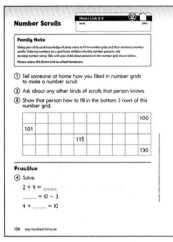

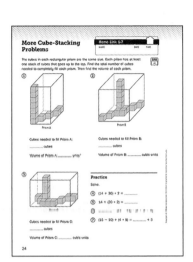

Sample first-grade *Home Link.*
© McGraw-Hill Education.

Sample fifth-grade *Home Link.*
© McGraw-Hill Education.

parents an opportunity to see what their child has done in math class. *Home Links* are available in both print and digital form.

In kindergarten through third grade, a Family Note will be found at the top of most *Home Link* pages. These notes appear less frequently in grades four through six, as older students tend to work more independently. Family Notes are written for adults at home and may include the following:

- Information about what was done in class that day
- Background about the mathematics involved in the Home Link
- Suggestions on how to further help your child

Occasionally, the Family Note comes as a separate page providing more detail about a topic with which parents may be unfamiliar.

Home Links are just one of many opportunities for practice that your child will have during the school year, so there's no need for a page full of repetitive computation problems. Most *Home Links* will include a few key problems that provide additional practice with the math involved in the main ideas of the day's lesson. In addition, many *Home Links* include a strip of practice exercises at the bottom of the page. These exercises often take the form of computation and facts practice, but they may include other types of practice in early grades. The purpose of the exercises is to provide another avenue for distributed practice with basic skills. Distributed practice—review in small doses spread out over time—is much more efficient and effective than more traditional approaches, which deliver practice in big chunks or masses.

The suggested method for completing *Home Links* varies by grade level. In kindergarten, first, and second grades, *Home Links* are designed to be done at home along with a parent, guardian, older sibling, or other adult. It is important for children to have a chance to discuss what they did in math class each day, so many *Home Links* involve prompts to explain a skill or concept to someone at home. In third grade, children transition to more independent work, while still sometimes working with someone at home. In fourth through sixth grades, *Home Links* are intended to be completed by students independently.

Many *Home Links* include a *My Reference Book* (G1–2) or *Student Reference Book* (G3–6) icon. These icons point children to reference materials they may find helpful in completing their homework.

 My Reference Book icon. © McGraw-Hill Education.

 Student Reference Book icon. © McGraw-Hill Education.

The icons include a page number or a range of page numbers that explain concepts on the Home Link page. Whenever they need further help on a particular topic, children and parents can refer to the reference book. Parents can also refer to a Family Letter for answers to the Home Link problems, and additional homework help can be found online at http://everydaymath.uchicago.edu/parents. In addition to the Family Note found on many *Home Links*, teachers may distribute Family Letters at the beginning of each unit, with information about upcoming lessons. Family Letters can also be found online at http://everydaymath.uchicago.edu/parents.

All Family Letters begin with a brief description of what your child will learn in the coming unit and may also include a description of the math tools your child will be using. Family Letters also always include the following sections:

- *Vocabulary.* This section lists some of the new vocabulary that your child will be exposed to in the unit. Most of these terms should be familiar to children by the end of the unit. Some of the vocabulary appearing on *Home Links* might only be mentioned in class; in these cases, children are not expected to know and use the actual term. Instead, they are simply to understand the concept and become familiar with the term.

- *Do-Anytime Activities.* This is a list of activities parents can "do anytime" with their child to reinforce the concepts they are learning.

- *Building Skills through Games.* This section describes selected games that your child will play throughout the unit. Playing these games at home is one way parents can give their child great opportunities for additional practice.

Second-grade Family Letter. © McGraw-Hill Education.

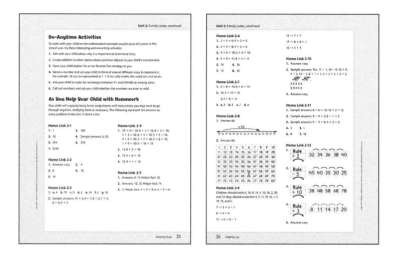

- *As You Help Your Child with Homework.* Here you will find answers to many of the *Home Links* problems for the unit.

Another *Everyday Mathematics* resource children can use at home is their *Everyday Mathematics* reference book. *My Reference Book* (*MRB*) is a resource for younger children which they can use to find out more about the mathematics learned in class.

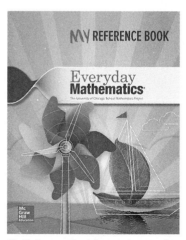

My Reference Book for grades 1 and 2.
© McGraw-Hill Education.

Introduced in the second half of first grade and used throughout second grade, the *MRB* is meant to be read with an adult, although many children learn to use the book on their own. In the *MRB*, children can look up and review explanations and find worked examples of the mathematics taught in class and directions for some of the more popular *Everyday Mathematics* games. You and your child can use the *MRB* together to find help with *Home Links*, to discuss the day's mathematics lesson, or simply to answer interesting mathematical questions that might arise.

In grades 3 through 6, children use a grade-specific *Student Reference Book* (*SRB*). In addition to the information included in *My Reference Book*, all *SRB*s include a glossary of mathematical terms, and some of the *SRB*s include real-world data that children use while solving problems in class.

Students use Real World Data from the *SRB* to complete the page. © McGraw-Hill Education.

Teachers may send the reference books home for children to use while doing their homework or they may ask you to use an electronic version. See section 4 for information about accessing reference books online.

Using a Calculator Is NOT Cheating

Calculators are one of the many valuable tools children learn to use in *Everyday Mathematics*. However, calculators are also the tool that seems to be the most controversial. Many parents, when hearing that their children use a calculator in math class, question whether using calculators will prevent children from learning how to compute independently. Fortunately, many educational researchers have wondered the same thing, and in recent years the effects of calculator use on children's mathematical learning has been well studied. The great majority of this research finds conclusively that using calculators appropriately can strengthen children's understanding and mastery of arithmetic, bolster good number sense, enhance problem-solving skills, and improve children's attitudes toward mathematics—and do all of this without interfering with the development of necessary paper-and-pencil calculation skills.

In *Everyday Mathematics,* calculators are used to help children learn mathematics. The authors find that as early as kindergarten they can be valuable for teaching a variety of skills. For example, in kindergarten children enter single-digit numbers on a calculator to represent quantities. *Everyday Mathematics* makes regular use of other calculator games throughout the program as well.

Everyday Mathematics uses calculator activities to help children develop number sense and flexibility with numbers. For example, in first and second grades, children do a "broken calculator" activity in which they pretend that one key of their calculator is broken. They are asked to show different ways of making a given number appear on the display without using the "broken" key. For example, if challenged to make 12 appear on the screen without using the 2 key, children come up with multiple representations for the number 12, such as $6 + 6$, 4×3, $1 + 1 + 1 + 1 + 1 + 1 + 1 + 1 + 1 + 1 + 1 + 1$, $13 - 1$, and so on.

Calculators also play an important role in *Everyday Mathematics* in helping children develop other mathematical skills and understandings. Children use calculators to count by 1s;

Broken Calculator problems.
© McGraw-Hill Education.

CAN YOU CHANGE THE 3 IN 4,349 TO A 0 WITHOUT CLEARING YOUR CALCULATOR?

I CAN SUBTRACT 300.

Calculators help children develop their understanding of place value.

to skip count by 2s, 5s, and 10s; to understand symbols such as +, −, =; and to model and solve number stories. This work is extended in later grades, when children use calculators not only to make calculations but also to explore new content such as measurement conversion and exponents and to experiment with the order of operations.

Beginning in first grade and continuing through fourth grade, children play a game called *Beat the Calculator* in which classmates race to solve basic math facts, one using a calculator and one solving the facts mentally. The goal is for the child solving the facts mentally to beat the child using the calculator, encouraging the development of quick mental skills, which is very different from depending on a calculator for basic facts.

In *Everyday Mathematics*, children learn to evaluate problems to decide when it is appropriate to use a calculator and when using mental arithmetic or paper and pencil is the better choice. Often this is based on how complicated a given problem is, but the choice can also depend on whether the problem calls for an exact answer or an estimate. Most children eventually acquire good judgment about when to use and when not to use calculators. Exercises aimed specifically at improving children's computational fluency, for which using a calculator would be inappropriate, are labeled with a no-calculator icon.

Like any other tool, calculators can be misused in ways that produce invalid results, and children must learn to check whether the answer they get using a calculator makes sense. Children might get a nonsensical result for several reasons:

Partial-Sums Addition

For Problems 1–3, make a ballpark estimate. Then solve the problem using partial-sums addition. Show your work. Use your estimate to check that your answer makes sense.

Unit

Example: 59 + 26 = ?
Ballpark estimate:
60 + 30 = 90

59
+ 26
70
15
85

① Ballpark estimate:

34
+ 71

② Ballpark estimate:

136
+ 157

③ Ballpark estimate:

122
+ 53

④ Solve one of the problems a different way. Explain your strategy.

158 one hundred fifty-eight

The no-calculator icon used in *Everyday Mathematics.* © McGraw-Hill Education.

- They neglected to clear the calculator before starting a new problem.
- They entered a number or operation incorrectly.
- They analyzed the problem incorrectly.

Even when solving a problem with a calculator, children learn to judge whether the answer they get is reasonable by asking whether it makes sense in terms of the original problem situation, just as they would if they solved the problem with paper and pencil or mental arithmetic.

As previously mentioned, the authors of *Everyday Mathematics* believe it is important for children to learn paper-and-pencil algorithms. However, once these algorithms are understood and mastered, having to perform them repeatedly can become tedious and dull and may lead children to dislike mathematics. So another benefit of using calculators is that they free children from having to repeat tasks that they already fully understand and from which they derive no real benefit. And calculators allow children to solve thought-provoking problems that demand computations that might otherwise be too difficult or time consuming, including problems from outside of mathematics class. By teaching children to use calculators wisely, *Everyday Mathematics* shifts the focus onto the problems themselves, rather than on carrying out algorithms.

Fifth-grade students solve the Paying Off the National Debt problem. Using a calculator allows them to focus on problem solving, rather than on the complex computations. © McGraw-Hill Education.

FURTHER READING

Campbell, P. F., & Stewart, E. L. (1993). Calculators and computers. In R. Jensen (Ed.), *Research ideas for the classroom: Early childhood mathematics*. New York: Macmillan.

Demana, F., & Leitzel, J. (1988). Establishing fundamental concepts through numerical problem solving. In A. F. Coxford & A. P. Shulte (Eds.), *The ideas of algebra, K–12* (1988 Yearbook). Reston, VA: National Council of Teachers of Mathematics.

Groves, S., & Stacey, K. (1998). Calculators in primary mathematics: Exploring numbers before teaching algorithms. In L. J. Morrow & M. J. Kenney (Eds.), *The teaching and learning of algorithms in school mathematics*. Reston, VA: National Council of Teachers of Mathematics.

Ellington, A. J. (2003). A meta-analysis of the effects of calculators on students' achievement and attitude levels in precollege mathematics classes. *Journal for Research in Mathematics Education, 34*(5), 433–463.

Hembree, R., & Dessart, D. J. (1992). Research on calculators in mathematics education. In J. T. Fey & C. R. Hirsch (Eds.), *Calculators in mathematics education* (1992 Yearbook). Reston VA: National Council of Teachers of Mathematics.

National Council of Teachers of Mathematics. (2015). *Position statement: Calculator use in elementary grades*. Reston, VA: Author. Retrieved from http://www.nctm.org/Standards-and-Positions/Position-Statements/Calculator-Use-in-Elementary-Grades/

National Council of Teachers of Mathematics. (2015). *Position statement: Strategic use of technology in teaching and learning mathematics*. Reston, VA: Author. Retrieved from www.nctm.org/Standards-and-Positions/Position-Statements/Strategic-Use-of-Technology-in-Teaching-and-Learning-Mathematics/

Smith, B. A. (1997). A meta-analysis of outcomes from the use of calculators in mathematics education. *Dissertation Abstracts International, 58*, 787A.

Waits, B. K., & Demana, F. (2000). Calculators in mathematics teaching and learning: Past, present, and future. In M. J. Burke & F. R. Curcio (Eds.), *Learning mathematics for a new century* (2000 Yearbook). Reston, VA: National Council of Teachers of Mathematics.

Fun and Games:
Math Is So Much More than (S)Kill and Drill

Most parents probably remember teaching their children to tie their shoes or ride a bike. Initially, these can be frustrating experiences for both parents and children because children rarely master either skill the first time they try. In fact, people rarely master anything the first time they encounter it, and mathematics is no different. Just as children have to practice tying their shoes and riding their bikes, they also need frequent practice spaced over time to master mathematical skills and concepts. But practice that comes solely in the form of drills can be monotonous and is often counterproductive because it turns children off to the subject. This is why *Everyday Mathematics* promotes mastery by providing repeated exposure to skills in a variety of different ways, not just as drill.

Another special feature of the *Everyday Mathematics* program is the way it uses games as an engaging way for children to practice mathematical skills. Games are an integral part of *Everyday Mathematics*, not an optional extra. Games

---

- make practice fun;
- can be played many times without problems being repeated because in most games numbers are generated randomly with dice, cards, or spinners;
- can be modified to meet the individual children's needs;
- provide a natural vehicle for children to discuss their mathematical thinking; and
- provide an important opportunity for teachers to informally assess skills and understandings.

The practice that comes from playing games has been shown to build number sense, fact recall, and procedural skills. Playing games also reinforces other skills and understandings, including calculator skills, place-value concepts, logic and reasoning, communication, and geometric intuition.

In *Everyday Mathematics*, games are not used as rewards or time-fillers. They are an essential part of the practice structure. The games played at all levels of the program were carefully designed to emphasize important mathematical content and provide meaningful practice of key skills and ideas. A major focus of games in kindergarten through third grade is basic facts. These games serve as a rich and engaging replacement for the traditional practice of drills, flash cards, and timed testing, meaning that children's mastery of basic facts depends in part on teachers and parents faithfully incorporating game play as recommended in the program. Skipping games at these levels deprives children of math facts practice they need.

Everyday Mathematics includes games at all grade levels not only for computation practice, but also for many other content areas, including geometry, measurement, and fractions. In third grade, for example, children begin developing deeper understandings of the concepts of area and perimeter. Area is the amount of surface inside a shape; perimeter is the distance around the boundary of a shape. These concepts are central to work in geometry

THE AREA IS 10, BUT THE PERIMETER IS 14. I WANT TO SCORE THE MOST POINTS, SO I'LL CHOOSE PERIMETER.

The Area and Perimeter Game

and measurement, but children frequently confuse them, often struggling with misconceptions many years after area and perimeter have been introduced. *Everyday Mathematics* addresses this difficulty in part with a game called *The Area and Perimeter Game*. Children draw one card showing a rectangle with a given length and width and a second card telling them whether to calculate the area, the perimeter, or to choose one or the other. The calculated measurement becomes the player's score for that hand. When a player draws a card allowing free choice, he or she determines which is larger—area or perimeter—and calculates that measurement, knowing that the larger measurement scores more points. By being called upon to compare and contrast available options, children both deepen their appreciation of the meaning of these key terms and practice calculating area and perimeter.

Children typically enjoy the games so much that they often ask to play them during their free time in class, and many teachers also make games available to be played at home. Most games use materials that can be commonly found at home, including decks of cards, dice, or small counters, such as pennies or dried beans. Teachers may provide printouts of special cards, spinners, or other materials needed for the remaining games. Some games use specially labeled dice, which can be easily replicated at home by writing numbers on small pieces of paper and taping them to the sides of a regular die. You will find descriptions of selected games in the *Family Letter* sent home prior to the beginning of a new unit. You

Die for *Multiplication Draw*, a game for practicing 2s, 5s, and 10s multiplication facts.

can also find directions for playing most of the major games in your child's grade in the *My Reference Book* or *Student Reference Book*, both of which are available in print and online.

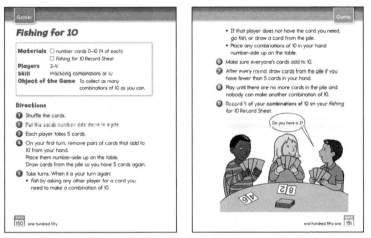

Directions for playing *Fishing for 10* in *My Reference Book*. © McGraw-Hill Education.

Many games can be easily altered to find the right balance between practice and challenge for your child. For example, think back to the game *Addition Top-It*, in which children flip over two number cards, find the sum of the numbers, and compare the sum with their opponent's sum. This game can be adapted in many ways, including these:

- For novice players, remove the larger number cards or use dice instead, as dice make it easier for younger children to count to find the sums.
- Provide more of a challenge by having each player turn over three number cards (or more) to add together.
- Include "wild cards" that allow players to pick any number they want whenever a wild card is drawn.
- Change the operation practiced by playing *Subtraction Top-It* or *Multiplication Top-It*.

Many other *Everyday Mathematics* games can be tailored in similar ways to provide your child with the most appropriate and engaging practice possible.

You may also wish to take advantage of the *Everyday Mathematics* games that can be played online. These digital versions of games provide children immediate and accurate feedback on each response they give and often come in both one- and two-player modes. If your child's school has purchased online access to the program, you can use the login information provided by your child's teacher to access online games via a home computer. After selecting your child's grade level, you will have the opportunity to choose games across three levels of difficulty: Skillbuilder, Grade Level, or Challenge. Note that games are sorted by number of players, allowing children to select from games that they can play alone or with a guardian, sibling, or friend.

If you are interested in practice "on-the-go," you can find a number of *Everyday Mathematics* apps by searching for "Everyday Mathematics" through your handheld device's app store. Popular number sense and facts practice games, such as *Top-It, Name That Number*, and *Beat the Computer* are available for a small price. Among the many mathematics apps that are available, you will be able to distinguish the genuine *Everyday Mathematics* apps by looking for those produced by McGraw-Hill School Education Group.

Finally, the quality of practice during game play depends largely on how you interact with your child while playing. It is important for children to explain occasionally how they found their answers, sharing and discussing both correct and incorrect ideas. This provides children with valuable practice articulating their mathematical thinking, and giving their own explanations will often lead them to self-correct incorrect answers. Use questions such as the following to encourage good mathematical thinking during game play:

- How did you figure out your answer?
- Can you say out loud how you thought about the problem in your head?
- Is there another way you could figure the answer out?

- Can you try that strategy on this next problem?
- Is there something you know that could help you to figure this problem out?
- If someone didn't know the answer to ___, how would you tell them to figure it out?

Questioning children in this way encourages them to think more deeply and keeps the focus on the mathematics, ensuring a high-quality practice experience.

Just as practice is important for learning to ride a bike or tie a shoe, practice is critical for mastering mathematical skills and concepts. The entertaining, meaningful, and varied options for practice provided by *Everyday Mathematics* allow children to master skills and concepts while also having fun. Games provide an ideal vehicle for this, particularly when they are thoughtfully adapted and children are engaged in discussion about their mathematical thinking.

GO ONLINE
http://everydaymath.uchicago.edu/about/understanding-em/games

What to Expect at Each Grade Level of *Everyday Mathematics*

At every grade level, the *Everyday Mathematics* curriculum is designed so that the vast majority of children will achieve a specific set of goals by the end of the year. This includes all of the content standards described in the Common Core State Standards for Mathematics.* This section is ordered by grade level and is meant to help you understand what mathematics your child will be expected to know and do at different times throughout the year, concluding with what can be expected by the end of the year. The descriptions include only the most important content at each grade level, not all of what is addressed over the school year. In addition to performance expectations, in this section you will also find Do-Anytime Activities that you can do with your child at home to support his or her learning.

If it appears your child is not progressing at the pace detailed in this section, keep in mind that children learn at different rates. Only rarely are all children in a class or grade at the same level at the same time for any given skill or concept. The expectations spelled out here indicate what the majority of children will do as they move through the year. If you continue to have concerns about your child's progress as time goes on or your child is progressing at a much faster rate, it is important to talk with your child's teacher, as he or she will assess your child regularly, if not daily, as well as at the conclusion of each unit.

For more detailed information about specific instruction in Operations with Whole Numbers, Math Facts, Fractions, Early Measurement, and Algebra, see "Why *Everyday Mathematics*? Because of How Key Content Is Taught."

---

* For a comprehensive listing of the Common Core State Standards for Mathematics, detailing what students should understand and be able to do at every grade, see www.corestandards.org/Math/.

KINDERGARTEN

NUMBERS AND COUNTING

Expectations

Early in the year, expect your child to count by 1s from 0 or 1 to at least 10. He or she will also correctly count sets of up to 10 objects, using one number word for each object counted (understanding one-to-one correspondence), and understand that the last number said is the total number in the set (the cardinal principle).

As the year progresses, kindergartners count on from numbers other than 0 or 1 and count up to higher numbers, learn to count by 10s, and count larger sets of objects, including objects that are not neatly arranged. Your child will also read and write numerals, first through 10 and then through at least 20, and use these numerals to represent quantities. And your child will start using his or her growing knowledge of numbers and quantities to compare sets of objects, as well as numerals, telling which is greater or less.

By the end of kindergarten, expect your child to count on by 1s from any number to at least 100; count by 10s to 100 (10, 20, 30, 40, 50, and so on, to 100); read and write numbers to at least 20; count and count out sets of up to 20 objects; and compare sets of objects and written numerals.

Do-Anytime Activities

- Count out loud together to pass the time while waiting, walking, driving, or cleaning up. Sometimes start from 1; sometimes start from other numbers. Count by 10s after your child learns this skill. Use funny voices (such as the big bad wolf or a little old lady) to keep counting activities interesting and playful.
- Find lots of different reasons to count together and to compare quantities. For example: *How many steps to the front door? Can you give me 15 pennies? Do we have more apples or more oranges?* Ask your child to record the numbers to help you both remember what you counted.
- Play a variety of card games like *Go Fish* and *Top-It* (also

known as *War*) to provide fun practice with numeral recognition, number comparisons, and other numeration skills. Standard playing cards are good to use because they help your child associate quantities (of hearts, clubs, and so on) with the corresponding numerals on the cards. Remove the face cards if desired, and encourage your child to invent new card games.

PLACE VALUE WITH WHOLE NUMBERS

Expectations

Early in the year, your child will focus on working with single-digit numbers and on the meaning of 0.

Midway through the year, kindergartners begin exploring the numbers from 10 to 19, with special attention to place value. Using hands-on materials, such as fingers on two people's hands or counters on double ten frames, your child will learn to represent these numbers concretely as 10 ones along with an additional number of ones.

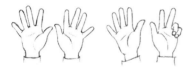

Your child will show 18 as 10 fingers and 8 more.

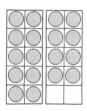

Your child will use a double ten frame to show 18 as 10 ones and 8 more ones.

By the end of kindergarten, expect your child to compose and decompose numbers from 10 to 19 into 10 ones and some more ones concretely (with materials or drawings) and symbolically (with equations).

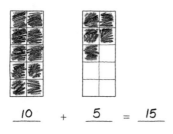

Your child will represent teen numbers (11 through 19) concretely on double ten frames and with a corresponding equation.

$$\underline{10} \quad + \quad \underline{5} \quad = \quad \underline{15}$$

Do-Anytime Activities

- Have your child show different teen numbers using ten fingers on your hands, as well as the correct number of fingers on his or her own hands. Talk about how the number is written with numerals (for example: 13). Later in the year, ask your child to write an equation to match the situation (e.g., 10 + 3 = 13).
- Play *Guess My Number* using place-value clues for numbers from 10 through 19. For example: *I'm thinking of the number that is 10 and 6 more.* Or, *I'm thinking of the number that goes with 10 to make 18.*

OPERATIONS WITH WHOLE NUMBERS

Expectations

Early in the year, expect your child to use fingers, objects, or drawings, or to act out situations to solve simple number stories involving addition and subtraction within 5. For example: *We have these 3 toy cars and those 2 toy trucks. How many vehicles do we have all together? How many will we have if we give one of the cars to our neighbor?*

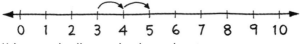

Using a number line to solve the number story.

As the year progresses, your child will solve number stories with slightly larger numbers (through at least 10) and may begin using less concrete strategies, such as using a number line or working from or remembering known combinations.

Your child will also learn to think about part-whole number relationships as he or she composes and decomposes numbers using manipulatives. For example, children find a variety of ways to make (compose) or take apart (decompose) a stack of 5 cubes using different numbers of cubes in two (or more) colors.

Your child will decompose and compose 5 in different ways. © McGraw-Hill Education.

By the end of kindergarten, expect your child to solve number stories and problems involving addition and subtraction within 10 using objects, drawings, or other strategies, and to decompose numbers less than or equal to 10 into pairs in more than one way.

Kindergarten Activity Card. © McGraw-Hill Education.

Do-Anytime Activities

- Tell your child number stories involving addition, subtraction, and part-whole relationships using familiar and interesting names and situations. Have him or her solve the problems and explain the strategy used. For example: *There were 5 kids at the bus stop this morning. Three of them were girls. How many were boys?* As the year progresses, increase the difficulty of the problems. You may also have your child tell number stories for you to solve.
- Play *Guess My Number* using addition and subtraction clues. For example: *My number equals 4 + 3. What is my number?*

What is another addition or subtraction number sentence that matches my number? Or, My number plus 2 equals 6. What is my number?

MATH FACTS

Expectations

Early in the year, your child will develop an understanding of the meanings of addition and subtraction. It is important for children to spend time focusing on this conceptual foundation first, before starting to learn particular addition or subtraction facts.

As part of this early conceptual work your child will begin developing strategies for addition and subtraction facts, such as "seeing" numbers in groups when various dot patterns are shown or understanding a pattern for identifying the number that is one more (+1) or one less (−1) than any given number.

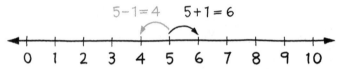

$$5-1=4 \qquad 5+1=6$$

Your child will come to understand that the number that is one more than a number (+1) comes right after the number when you count, and the number that is one less (−1) comes right before it.

As the year progresses, your child will also begin to investigate number pairs that add to 10, laying the groundwork for facts work in later grades.

By the end kindergarten, your child will draw on the conceptual understanding and strategies he or she has developed to add and subtract fluently within 5.

Do-Anytime Activities

- Customize two dice by labeling one with 1, 1, 2, 2, 3, 3, and the other with 0, 0, 1, 1, 2, 2. (You can put masking tape or dot stickers over the sides of existing dice, or you can mark up small wooden blocks or cubes.) Have your child roll both dice and find the sum as quickly as he or she can. Encourage your child to use strategies that are quick and accurate and that make intuitive sense to him or her.

- Customize two dice by labeling both with 0, 1, 2, 3, 4, 5. Have your child roll both dice and find the difference between the larger number and smaller number as quickly as he or she can. Encourage your child to use strategies that are quick and accurate and that make intuitive sense to him or her.
- Play *Guess My Number* using the making-10 strategy and +1/−1 clues. For example: *I'm thinking of the number that makes 10 when it is added to 4. What's my number?* Or: *I'm thinking of two numbers: one equals 7 + 1 and one equals 7 − 1. What are my numbers?*

DATA

Expectations

Early in the year, your child will sort and classify objects in various ways, such as by size, color, or shape. He or she will also use categorization to create and use many simple data displays in school, such as a graph showing birthday months and ages of the children in the class.

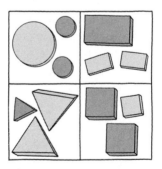

Blocks sorted into shape categories

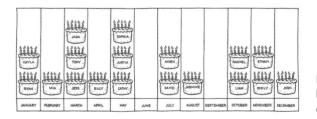

Early kindergarten data work

As the year progresses, your child will create and use more complicated data displays, paying attention to counting and comparing the number of items in each category when sorting objects and making graphs.

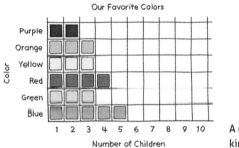

Our Favorite Colors

A more complicated kindergarten data display

By the end of kindergarten, expect your child to classify objects into given categories, count the number of objects in each category, and sort the categories by count.

Do-Anytime Activities

- Encourage your child to help out with sorting tasks around the house, such as sorting laundry, putting away silverware, or organizing pantry shelves. Have him or her count, compare, and order the number of objects of particular types. For example: *How many spoons do we have? How many forks? How many knives? Do we have more forks or more spoons? Which utensil do we have the most of?*
- Have your child collect and organize data based on his or her interests and surroundings. For example: *How many shells of each shape do you have in your beach collection? Let's line them up by shape and count them. Can you think of a way to show it with a picture or chart?*

MEASUREMENT

Expectations

Early in the year, your child will describe and directly compare the lengths of objects. He or she will talk about what *length* refers to as a concept and look for objects that are longer than, shorter than, and the same length as other objects, practicing using length terms such as *long, longer, short, shorter, tall,* and *taller.* Your child will explore and discover why it is important to line up ends when comparing lengths.

Around midyear, your child will begin to explore other measurable attributes of objects, such as weight and capacity. He or she will learn to talk about what each dimension refers to and make direct comparisons by using a pan balance (for weight) and by pouring (for capacity). Children now begin to describe their findings using terms such as *heavy, heavier, light, lighter, holds more,* and *holds less.*

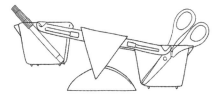

Which is *heavier?*

By the end of kindergarten, expect your child to describe measurable attributes of objects, including describing several measurable attributes of a single object. Also expect him or her to directly compare objects along a particular dimension, such as length or weight.

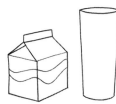

Which *holds more* (has more capacity)?

Do-Anytime Activities

- Have your child describe the size of a given object in as much detail as he or she can. For example: *Tell me about the height, weight, and capacity of this bowl.*

- Engage your child in comparing and ordering objects along a particular size dimension in real-life situations. For example, ask him or her to stack books on the coffee table with the heaviest one on the bottom or to line up books on the bookshelf from tallest to shortest. Model and encourage the use of precise size language (e.g., *longer* or *heavier* instead of *bigger*).

GEOMETRY

Expectations

Early in the year, your child will explore a large variety of shapes, using them to practice identifying, describing, and comparing different characteristics and features, such as straight, curved, skinny, and so on. He or she will also begin to use positional terms, such as *above, below,* and *next to.*

Soon after, your child will start using these characteristics and features to identify, describe, and compare circles, triangles, and rectangles (including squares). Children will be exposed to these basic shapes in a range of sizes, proportions, and orientations, and be encouraged to identify and name them in their surroundings. As the year progresses, your child will draw these shapes and model them with other materials, such as string, clay, and toothpicks.

Around midyear, your child will compose shapes to form larger shapes, such as joining two triangles to form a rectangle or to form a larger triangle.

Later in the year, children begin analyzing and comparing the attributes of 3-dimensional shapes, including spheres, cylinders, cones, and rectangular prisms (including cubes), learning to identify shapes as 2-dimensional or 3-dimensional.

By the end of kindergarten, expect your child to name, describe, and compare basic 2-dimensional and 3-dimensional shapes, model and draw shapes, and combine shapes to make other shapes. Also expect him or her to use a range of positional terms to describe the relative positions of objects.

Do-Anytime Activities

- Go on Shape Scavenger Hunts to look for geometric shapes in your surroundings (indoors and out). Have your child name and describe the shapes he or she finds and also describe their location relative to other objects. Bring a clipboard and paper and encourage your child to sketch some of the shapes.
- Using building toys like blocks, interlocking bricks, and ball-and-stick sets, encourage your child to describe the shapes of the individual pieces, as well as the shapes he or she makes from combining them.
- Encourage your child to assemble puzzles. Discuss how the pieces fit together to form new shapes.

ALGEBRA

Expectations

Early in the year, your child will represent problems and number stories with fingers, objects, pictures, and words. For example: *Davon has 3 marbles, and Ellie has 2 marbles. How many marbles do they have all together.*

 Representing the number story with a picture.

As the year progresses, children learn about the +, −, and = signs and begin to represent number stories symbolically with equations. For example: *Davon lost 2 of the marbles. 5 marbles − 2 marbles = 3 marbles.* Your child will also work to solve and represent number stories with missing numbers. For example: *Maya had 2 markers. She found more markers. Then she had 4 markers. How many more markers did Maya find?*

By the end of kindergarten, expect your child to use objects, pictures, words, numbers, and symbols to represent problems and number stories involving addition and subtraction.

Do-Anytime Activities

- Have your child use objects or pictures to represent number stories that you tell or he or she tells. Later in the year, have your child represent the number stories with equations.

NUMBERS AND COUNTING

Expectations

Early in the year, expect your child to count by 1s to 100 using a number line or number grid.

Number line

−9	−8	−7	−6	−5	−4	−3	−2	−1	0
1	2	3	4	5	6	7	8	9	10
11	12	13	14	15	16	17	18	19	20
21	22	23	24	25	26	27	28	29	30
31	32	33	34	35	36	37	38	39	40
41	42	43	44	45	46	47	48	49	50
51	52	53	54	55	56	57	58	59	60
61	62	63	64	65	66	67	68	69	70
71	72	73	74	75	76	77	78	79	80
81	82	83	84	85	86	87	88	89	90
91	92	93	94	95	96	97	98	99	100
101	102	103	104	105	106	107	108	109	110

Number grid

He or she will also count and represent numbers of objects up to 20 with written numerals.

As the year progresses, your child will begin skip counting by 2s (2, 4, 6, 8, 10, and so on), 5s (5, 10, 15, 20, 25, and so on), and 10s (10, 20, 30, 40, and so on) within 100. Your child will also compare the value of two numbers up to 20. For example: *8 is less than 12.* (See also the expectations under first-grade "Place Value with Whole Numbers.")

By the end of grade 1, your child can be expected to count on from any number to at least 120; read and write numbers to at least 120; and count the objects in a collection, representing the quantity with a numeral up to at least 120.

Do-Anytime Activities

- Count orally by 1s, 2s, 5s, and 10s while doing chores or riding in the car or on a bus. Sometimes count down (or back), for example, 45, 40, 35, 30, and so on. Later in the year, try starting at different numbers. For example, count up by 10s starting at 4: 4, 14, 24, 34, and so on.

- Count numbers of objects around your home or while shopping. Have your child keep track using tally marks and then record the total number. For example, count the number of spoons in your kitchen or the number of boxes in your pantry. Begin with small quantities, and as the year progresses, use larger sets of objects.

PLACE VALUE WITH WHOLE NUMBERS
Expectations

Midway through the year, first graders are introduced to place value. Your child will begin to identify the 2-digit number represented by base-10 blocks and, soon after, tell the value of each digit in a 2-digit number. At this time, your child will use what he or she knows about place value to compare numbers to 20 using the symbols >, <, and =. For example: *14 > 11*. (See also the expectations under first-grade "Numbers and Counting.")

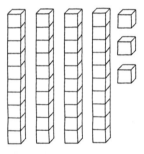

43 is composed of 4 tens, or 40, and 3 ones, or 3.

By the end of grade 1, your child can be expected to understand place value, including grouping by tens and ones, as well as to compare and order 2-digit numbers using >, =, and <.

Do-Anytime Activities

- Look for 2-digit numbers in and around your home or neighborhood. Ask your child to tell you how much each digit is worth. Later in the year, challenge your child to do the same with 3-digit numbers.

- Ask your child to compare 2-digit (and later 3-digit) numbers using <, >, and =.
- Have your child create the largest and smallest 2-digit (and later 3-digit) number when given 2 (or 3) digits. For example, given the digits 3 and 8, the largest 2-digit number is 83 and the smallest is 38.
- Create and solve puzzles from pieces of a number grid in which most of the numbers are missing.

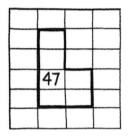

Fill in the missing numbers in the number-grid puzzle.

OPERATIONS WITH WHOLE NUMBERS

Expectations

Early in the year, expect your child to solve simple number stories involving addition and subtraction within 10. For example: *Madison has 2 red beads and 5 blue beads. How many beads does she have in all?*

As the year progresses, your child will begin to solve number stories with three addends. For example: *Madison has 2 red beads, 5 blue beads, and 5 yellow beads. How many beads does she have in all?* He or she will also use a number grid to find 10 more and 10 less than a number.

−9	−8	−7	−6	−5	−4	−3	−2	−1	0
1	2	3	4	5	6	7	8	9	10
11	12	13	14	15	16	17	18	19	20
21	22	(23)	24	25	26	27	28	29	30
31	32	33	34	35	36	37	38	39	40
41	42	43	44	45	46	47	48	49	50

10 more than 23 is 33.
10 less than 23 is 13.

Midway through the year, your child will begin adding 2-digit and 1-digit numbers using tools, such as a number grid or base-10 blocks, and find the difference between 2-digit multiples of 10 using tools.

−9	−8	−7	−6	−5	−4	−3	−2	−1	0
1	2	3	4	5	6	7	8	9	10
11	12	13	14	15	16	17	18	19	20
21	22	23	24	25	26	27	28	29	30
31	32	33	34	35	36	37	38	39	40
41	42	43	44	45	46	47	48	49	50
51	52	53	54	55	56	57	58	59	60
61	62	63	64	65	66	67	68	69	70
71	72	73	74	75	76	77	78	79	80
81	82	83	84	85	86	87	88	89	90
91	92	93	94	95	96	97	98	99	100
101	102	103	104	105	106	107	108	109	110
111	112	113	114	115	116	117	118	119	120

Number grid for adding and subtracting 2-digit and 1-digit numbers

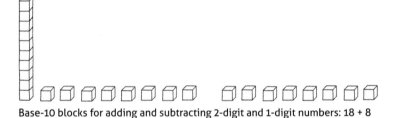

Base-10 blocks for adding and subtracting 2-digit and 1-digit numbers: 18 + 8

By the end of grade 1, expect your child to represent and solve number stories and problems involving addition and subtraction (including those with three addends), use tools or strategies to add and subtract within 100, and mentally find 10 more and 10 less than a given 2-digit number.

Do-Anytime Activities

- Tell your child number stories about everyday events. Have him or her solve them and explain the strategy used. For example: *I see 4 red cars and 6 black cars in the parking lot. How many cars do I see altogether?* As the year progresses, increase the difficulty of the problems. You may also wish to have your child tell number stories for you to solve.

- Select two numbers between 1 and 100. Ask your child to find their sum and their difference using tools such as a number grid. Early in the year, add 2-digit numbers to 1-digit numbers and multiples of 10, and subtract 2-digit multiples of 10. Later in the year, your child can practice adding and subtracting any 2-digit numbers.
- Name a 2-digit number. Ask your child to find 10 more and 10 less. Early in the year, allow your child to use a number grid, as needed, but work toward completing this task mentally as the year progresses.

MATH FACTS

Expectations

Early in the year, children use addition and subtraction within 10 to solve simple number stories and find pairs of numbers that add to 10. As the year progresses, expect your child to find and record facts within 10, including combinations of 10 (e.g., 4 + 6, 5 + 5, 8 +2, and so on) and doubles facts (e.g., 3 + 3, 7 + 7, 9 + 9, and so on). Later your child will use these facts to help solve other addition and subtraction facts within 20 using strategies like the near-doubles and making-10 strategies. (For more on these strategies, see "There's Nothing 'Basic' about Mastering Basic Facts.")

Example of a near-doubles strategy

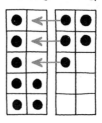

This Quick Look shows the making-10 strategy.

$7 + 5 = 7 + 3 + 2 = 10 + 2 = 12$

By the end of grade 1, your child can be expected to demonstrate fluency with addition and subtraction facts within 10 and to use strategies to solve addition and subtraction facts within 20.

Do-Anytime Activities

- Select a number less than 10. Have your child name the number needed to make a sum of 10. For example, if you say 6, your child should say 4.
- Roll a die. Have your child double the number shown and name the sum. For example, if your child rolls a 5, he or she will say 10. As the year progresses, use two dice. For example, if your child rolls a 3 and a 6, he or she will say 18 (the double of the 9 shown on the dice).
- Have your child show you how to use the making-10 strategy to solve 8 + 3. Have him or her suggest other facts that could be solved using this strategy.
- Have your child show you how to use the near-doubles strategy to solve 8 + 7. Have him or her suggest other facts that could be solved using this strategy.
- Once your child has received Fact Triangles from school, use them to help practice addition and subtraction facts.
- Play fact-related games found in *My Reference Book* or online in the Student Learning Center.

FRACTIONS

Expectations

First-grade work with fractions does not begin until late in the year, when children begin by dividing circles and squares into two equal parts. Soon after this, expect your child to partition shapes into two and four equal shares and name the shares in a variety of ways.

By the end of grade 1, expect your child to partition shapes into two and four equal shares and describe the shares and the whole using fraction words.

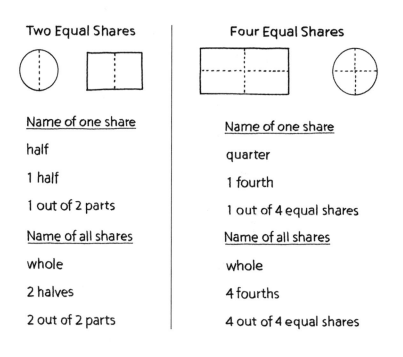

Two Equal Shares	Four Equal Shares
Name of one share	Name of one share
half	quarter
1 half	1 fourth
1 out of 2 parts	1 out of 4 equal shares
Name of all shares	Name of all shares
whole	whole
2 halves	4 fourths
2 out of 2 parts	4 out of 4 equal shares

Do-Anytime Activities

- Draw a picture of a rectangular cake, a circular pizza, or a similar food. Discuss ways to cut the food to feed 2 (or 4) people so that each person gets an equal share. Ask your child to share various names for each share and for the whole.
- Divide the same shape into 2 and 4 equal parts. Discuss which parts are larger.

DATA

Expectations

Early in the year, your child will be able to organize data in a tally chart and answer simple questions about a tally chart.

Right or Left Handed ?	
Right handed	⊬⊬ ⊬⊬ ////
Left handed	⊬⊬ ////

As the year progresses, children begin answering questions about bar graphs.

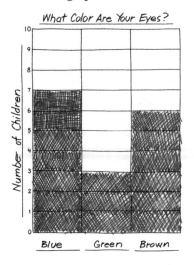

How many more children have blue eyes than green eyes?

By the end of grade 1, your child can be expected to collect and organize data, represent data using tally charts and bar graphs, and ask and answer questions about data in tally charts and bar graphs. In first grade, work with data is limited to three categories, so, for example, a survey of favorite types of ice cream flavors would ask only about chocolate, vanilla, and strawberry.

Do-Anytime Activities

- Collect data by asking questions about topics that will interest your child. For example: *What is your favorite type of pizza—meat, cheese, or vegetable? Which pet do you like best—dog, cat, or fish?*
- Collect data by making observations: For example: *How many people have brown hair, blond hair, or red hair? What is the weather today—sunny, rainy, or cloudy?*
- Organize data in tally charts and in bar graphs. Ask your child questions about the data. Then have your child formulate questions about the data to ask you.

MEASUREMENT

Expectations

Early in the year, children begin work with length measurement. At that time, expect your child to directly compare and order objects by length. He or she will also learn to measure the length of an object with multiple paper clips or pencils.

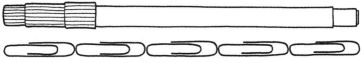

This marker is 5 paper clips long.

As the year progresses, your child will be introduced to time and can be expected to tell time to the hour on an hour-hand-only clock. Soon after this, your child will learn to indirectly compare the lengths of two objects using a third object. Your child will also tell time to the hour using an analog clock with both hour and minute hands.

An hour-hand-only clock showing about 3 o'clock

The gray bookshelf is taller than the length of this string, and the white bookshelf is shorter than the length of this string. So the gray bookshelf is taller than the white bookshelf.

By the end of grade 1, expect your child to compare and order objects by length; to measure length using one same-sized, non-standard unit; and to tell and write time to the hour and half hour on both analog and digital clocks.

Do-Anytime Activities

- Measure small objects in your home using paper clips. For example, you might measure the length of a cooking utensil, the width of a small table, or the length of a book. Early in the year, work with your child to place the paper clips end to end, without gaps or overlaps. As the year progresses, have your child measure the same things using only one paper clip by moving it along the object being measured.
- Ask your child to order a group of items in your home from shortest to longest. Have him or her explain how to compare the lengths of objects indirectly.
- Ask your child to tell time to the hour (and later the half hour) using an analog clock. Later in the year, have your child write time to the hour and half hour in digital notation.

GEOMETRY

Expectations

Early in the year, your child will be able to recognize sides and vertices (corners) of 2-dimensional shapes, as well as to draw and name 2-dimensional shapes. He or she will also combine shapes to make designs.

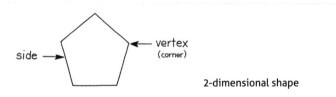

side →

← vertex (corner)

2-dimensional shape

As the year progresses, your child will be able to name the defining characteristics of 2- and 3-dimensional shapes, being able to answer, for example, *What characteristics make an object a triangle?*

Defining Attributes	Nondefining Attributes
All triangles ...	Some triangles and other shapes might ...
• have exactly 3 vertices.	• be different sizes.
• have exactly 3 straight sides.	• be different colors.
• are closed.	• have different patterns.
	• point in different directions.

He or she will also compose new 2-dimensional shapes from given 2-dimensional shapes.

By the end of grade 1, your child can be expected to distinguish between attributes that define shapes (e.g., four sides, no vertices [or corners]) and those that do not define shapes (e.g., red, large, small). He or she will combine shapes to form more complex shapes.

Do-Anytime Activities

- Look for geometric shapes in everyday objects around the house, at the market, in architectural features, and on street signs. Name the shapes using their geometric names, and have your child share defining attributes of the shapes.
- Construct polygons (closed, straight-sided, flat shapes) using drinking straws and twist ties from plastic storage bags.

Small-diameter straws, such as coffee stirrers, are easier to use and cut into 4- and 6-inch pieces. If only larger straws are available, fold the ends of the twist ties to fit tighter. To build polygons, put a twist tie into one end of at least three straws, using the ties to connect the straws.

- Make 3-dimensional shapes from straws and twist ties. To do this, put two twist ties (or one folded twist tie) into the ends of straws so that each end can be connected to two other straws.
- Encourage your child to build with blocks. Discuss how the pieces fit together in different ways to form new shapes.

ALGEBRA

Expectations

Early in the year, expect your child to represent problems and number stories with objects and pictures. For example: *Benny has 2 sisters and 3 brothers. How many siblings does Benny have in all?*

Representing the number story with a picture.

Your child will also observe that adding two numbers in a different order results in the same sum. For example: *4 + 9 is the same as 9 + 4.* He or she will be able to find the number of hops between two numbers on a number line or number grid.

0 1 2 3 4 5 6 7 ⑧ 9 10 11 12 13 14 15

How many hops from 8 to 12?

As the year progresses, your child will represent problems and number stories with number models. For example: a number model that represents the number story about Benny's siblings is 2 + 3 = 5.

Your child will apply the Commutative and Associative Properties of Addition to solve problems:

Commutative Property of Addition: If I know 3 + 6 = 9, then I also know 6 + 3 = 9.

Associative Property of Addition: (2 + 5) + 4 = 2 + (5 + 4), because 7 + 4 = 2 + 9.

Your child will also complete "What's My Rule?" problems, finding the rule relating two numbers and describing the relationship with a number sentence. (For examples of "What's My Rule?" tables, see "Algebra in Kindergarten?")

By the end of grade 1, expect your child to use objects, pictures, numbers, symbols, and words to represent problems and number stories involving addition and subtraction; find the unknown number in addition and subtraction equations, such as $8 + ? = 10$, $4 = 8 - \square$, and $3 + 3 = \_\_$; and understand properties of addition and the relationship between addition and subtraction.

Do-Anytime Activities

- Name two numbers less than 10. Have your child use a number line to tell how many hops are between the two numbers. Later in the year, use numbers up to 20.
- When practicing number stories, as discussed in "Moving On Up . . . Operations with Bigger Numbers," have your child represent the number stories with objects or by drawing pictures. Later in the year, have your child represent the number stories with number models.
- Have your child tell number stories that fit different equations, such as $6 + 6 = 12$ or $17 - 9 = 8$.
- Have your child explain to you why $3 + 2$ gives the same answer as $2 + 3$.
- Have your child explain how to use addition to solve subtraction problems. For example: *Which addition fact can help solve 12 − 9?* Expect your child to explain that knowing $9 + 3 = 12$ or $3 + 9 = 12$ helps them know $12 - 9 = 3$.
- Encourage your child to solve problems with $\_\_$ or $\square$ representing the unknown number. For example: $3 + \square = 5$ or $\_\_ - 7 = 13$.

NUMBERS AND COUNTING

Expectations

Early in the year, expect your child to count by 1s past 120, skip count by 5s (5, 10, 15, 20, 25, and so on), and skip count by 10s (10, 20, 30, 40, and so on) to at least 200. He or she will read and write numbers to at least 120 using base-10 numerals and to at least 10 using number names:

Base-10 numerals: 1, 2, 3, 4, . . . 10, 11, and so on
Number names: *One, two, three, . . . ten, eleven,* and so on

As the year progresses, children will skip count by 5s, 10s, and 100s within 1000. Your child will begin writing numbers to 20 using number names and writing numbers in expanded form. (See also the expectations under second-grade "Place Value with Whole Numbers.")

Expanded form: 357 = 300 + 50 + 7

By the end of grade 2, your child can be expected to be able to count on from any number by 1s, 5s, 10s, and 100s to 1000, and read and write numbers within 1000 using base-10 numerals, number names, and expanded form.

Do-Anytime Activities

- Count orally by 1s, 5s, 10s, and 100s while doing chores or while in the car or on a bus. Sometimes count down (or back). For example: 85, 80, 75, 70, 65, and so on. Later in the year, try starting at different numbers. For example: Count up by 10s starting at 7: 17, 27, 37, 47, and so on.
- Ask your child to name any 3-digit number and have him or her write a number model to show the expanded form (e.g., 694 = 600 + 90 + 4) and write the number name (e.g., six hundred ninety-four).

PLACE VALUE WITH WHOLE NUMBERS

Expectations

Early in the year, children will begin making exchanges between ones and tens and tens and hundreds. Your child will learn that 10 ones is equal to 1 ten and that 10 tens is equal to 1 hundred.

As the year progresses, your child will explore the values of digits by representing 3-digit numbers with base-10 blocks and expanded form. (See also the expectations under second-grade "Numbers and Counting.")

$$200 + 40 + 8 = 248$$

Your child will use place value and expanded form to compare two 3-digit numbers with nonzero digits:

347 = 300 + 30 + 7
325 = 300 + 20 + 5
The hundreds are the same, but 30 is more than 20, so 337 is larger than 325

By the end of grade 2, your child can be expected to understand place value, including grouping by ones, tens, and hundreds, and can be expected to compare and order 3-digit numbers using place value and >, =, and < symbols.

Do-Anytime Activities

- Use money to reinforce place-value exchanges. Make a pile of 20 pennies, 15 dimes, and a $1 bill. Have your child roll a die and take that number of pennies from the pile, trading 10 pennies for 1 dime and 10 dimes for 1 dollar when possible. Frequently ask your child how much money he or she has now.

- Ask your child to use expanded form to compare 3-digit numbers. Have him or her explain why one number is larger or smaller than another.
- Have your child create the largest and smallest 3-digit number when given 3 digits. For example, given the digits 5, 3, and 7, the largest 3-digit number is 753, and the smallest 3-digit number is 357.
- Ask your child to tell you the value of a digit in any 3-digit number. In 426, for example, the 4 is worth 400, the 2 is worth 20, and the 6 is worth 6.

OPERATIONS WITH WHOLE NUMBERS
Expectations

Early in the year, expect your child to write and solve simple number stories involving addition and subtraction using a number grid, number line, or counters. For example: *Nour has 12 crayons. While cleaning his room, he found 14 more crayons. How many crayons does Nour have now?*

−9	−8	−7	−6	−5	−4	−3	−2	−1	0
1	2	3	4	5	6	7	8	9	10
11	12	13	14	15	16	17	18	19	20
21	22	23	24	25	26	27	28	29	30
31	32	33	34	35	36	37	38	39	40
41	42	43	44	45	46	47	48	49	50
51	52	53	54	55	56	57	58	59	60
61	62	63	64	65	66	67	68	69	70
71	72	73	74	75	76	77	78	79	80
81	82	83	84	85	86	87	88	89	90
91	92	93	94	95	96	97	98	99	100
101	102	103	104	105	106	107	108	109	110
111	112	113	114	115	116	117	118	119	120

As the year progresses, children begin to mentally add and subtract 10 and 100 from any number. Your child will also begin to solve number stories involving addition and subtraction within

100 using strategies based on place value. He or she will continue to use a number grid, number line, or open number line to solve problems with larger numbers within 1000.

COMBINING 10S AND 1S	COUNTING UP FOR SUBTRACTION
26 + 74	76 − 34
20 + 70 = 90	34 + **40** = 74
6 + 4 = 10	74 + **2** = 76
90 + 10 = 100	40 + 2 = 42

Your child will use addition and subtraction strategies based on place value to solve problems.

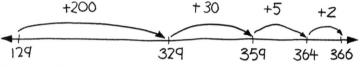

Your child will also use an open number line to solve problems.

Your child will begin to solve two-step number stories in second grade. For example: *Melissa bought 23 blue balloons and 13 white balloons for a party. During the party 7 balloons blew away. How many balloons did Melissa have left?*

Late in the school year, your child will begin arranging objects in rectangular arrays in preparation for the multiplication work that begins in grade 3. He or she will write addition number models using equal addends to express the number of objects in the array.

By the end of grade 2, expect your child to represent and solve 1- and 2-step number stories and problems involving addition and subtraction within 100, to use tools or strategies to add and subtract within 1000, and to mentally add and subtract 10 and 100 from a given 3-digit number. Your child will be able to explain why the chosen strategies work. Expect him or her to add up to four 2-digit numbers and to be able to find the total number of objects in a rectangular array and write an addition number model for the array.

Addition number model:

3 + 3 + 3 + 3 = 12

Do-Anytime Activities

- Tell your child number stories about everyday events. Have him or her solve them and explain the strategy used. For example: *I see 27 blue cars, 12 black cars, and 8 red cars in the parking lot. How many cars do I see altogether?* As the year progresses, increase the difficulty of the problems. You may also wish to have your child tell number stories for you to solve.

- Select two numbers between 1 and 100. Ask your child to use strategies to find their sum and their difference. Early in the year, add 2-digit numbers to multiples of 10 and subtract 2-digit multiples of 10 (23 + 30, 45 − 20, and so on). Later in the year, your child can practice adding and subtracting any 2-digit numbers.

- Name a 3-digit number. Ask your child to mentally find 10 more, 10 less, 100 more, and 100 less.

MATH FACTS

Expectations

Early in the year, your child will revisit addition and subtraction fact strategies and use known addition facts (doubles and sums of ten) to solve more difficult addition facts.

Example of near-doubles strategy

Example of making-10 strategy

Throughout grade 2, your child will discuss and practice different strategies to solve addition and subtraction facts.

By the end of grade 2, he or she can be expected to fluently add and subtract within 20 using mental strategies and know from memory all addition facts through 9 + 9.

Do-Anytime Activities

- Make sure your child knows the doubles facts (2 + 2, 3 + 3, 4 + 4, and so on) and combinations-of-10 facts (2 + 8, 3 + 7, 4 + 6, and so on). If needed, practice them in fun ways. For example, name a number and ask your child to double it. Or, name a number less than 10, and have your child name a number that added to your number equals 10.
- Take turns with your child creating and solving addition and subtraction number stories using facts.
- Play fact-related games found in *My Reference Book* or online in the Student Learning Center.

FRACTIONS

Expectations

Second-grade work with fractions does not begin until fairly late in the year, with children partitioning circles and rectangles into two, three, or four equal shares. Your child will begin describing the equal shares and the whole using word or number-word notation (two halves, 3-fourths, 1-third, three thirds, and so on). Your child will also begin to explore the idea that equal shares of the same-sized whole need not have the same shape.

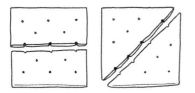

The sandwiches are the same size. Each sandwich shows 2 halves. The halves on one sandwich are rectangles and the halves on the other sandwich are triangles.

By the end of year, expect your child to partition circles and rectangles into two, three, and four equal shares, and describe the shares using the words *halves, half of, thirds, third of, fourths,* and so on. Expect your child to describe the whole as *two halves, three thirds,* and *four fourths.* You can also expect your child to recognize that equal shares of a shape don't have to have the same shape.

Do-Anytime Activities

- With your child, discuss ways to share food to feed 2, 3, or 4 people so that each person gets an equal share. Ask your child to describe each share and the whole using the words *halves, half of, thirds, third of,* and so on.
- Divide the same shape into 2, 3, and 4 equal parts in various ways and have your child describe the parts using fraction words.

DATA

Expectations

Beginning in the middle of the year, your child will be able to draw a picture graph to represent data in a tally chart.

Books Read in Mrs. Spring's Class

Name	Number of Books Read
Melissa	卌
Jose	
Beth	////
Justin	卌 ///
Sam	卌

Books Read in Mrs. Spring's Class

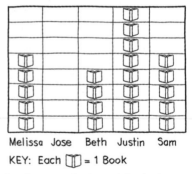

KEY: Each 📖 = 1 Book

Your child will create picture graphs using data organized on a tally chart.

As the year progresses, your child will learn to answer simple questions about data in a picture graph and bar graph.

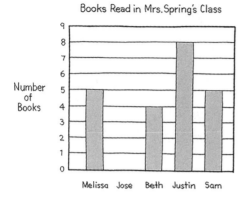

Books Read in Mrs. Spring's Class

How many more books did Justin read than Beth? How many books were read in all?

By the end of grade 2, your child can be expected to gather data, organize data, and represent data using bar graphs and picture graphs. Second-grade work with data for bar graphs and picture graphs is limited to four categories. For example, in a survey of favorite types of ice cream, flavor choices might be only chocolate, vanilla, peppermint, and strawberry. Children can also be expected to gather measurement data to the nearest whole unit (inches, centimeters, feet, yards, and meters), organize the data, represent it on a line plot, and be able to answer questions about the information in the data displays.

Height of Children in Mr. Adam's Class

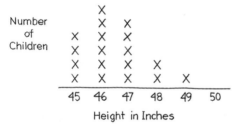

Your child will answer questions about information in line plots such as the following:

- What does it mean if there are a lot of Xs above a number?
- What does it mean if there are no Xs above a number?
- How many Xs are there above ___?
- How many more Xs are there above ___ than ___?
- What does that tell you? How many Xs are there all together?

- Collect data by asking questions of your child's choosing. For example: *What is your favorite type of pizza—sausage, pepperoni, cheese, or vegetable? Which pet do you like best— dog, cat, bird, or fish?*
- Collect data by taking measurements: For example: *How tall are the people in our family? How far can our family members jump?*
- Work with your child to organize data in bar graphs, picture graphs, and line plots. Ask your child questions about the data. Then have your child formulate questions about the data to ask you.

MEASUREMENT

Expectations

Early in the year, expect your child to tell time to the nearest hour (that it is about 2 o'clock, for example) using digital and analog clocks. He or she will also solve problems involving pennies and dimes. Your child will learn to measure the length of an object using both inches and centimeters.

As the year progresses, second graders are introduced to telling time to the nearest 5 minutes. At this point, expect your child to be able to tell time to the nearest half hour using an analog clock and distinguish between A.M. and P.M. Your child will also solve problems involving quarters, dimes, nickels, and pennies, as well as problems that involve lengths given in the same unit (inches, centimeters, feet, yards, or meters).

Xavier's arm span - 59 in

Lena's arm span - 42 in

difference

Xavier's arm span is 17 inches longer than Lena's.

Your child will be introduced to "personal measures" and will use them to make length estimates.

By the end of grade 2, expect your child to select appropriate tools to measure the length of an object using two different units (inches, centimeters, feet, yards, or meters) and describe how the two measurements relate to the size of the unit. He or she will measure to determine how much longer one object is than another and solve number stories involving measurements with the same units. Your child will estimate and measure lengths using inches, centimeters, feet, yards, and meters. Expect him or her to tell time to the nearest five minutes and distinguish between A.M.

and P.M. Your child will also solve problems involving coins and bills, and read and write money amounts.

Do-Anytime Activities

- Measure objects in your home using 6- or 12-inch rulers, tape measures, yard sticks, or meter sticks. For example, you might measure the length of the kitchen table, the width of a countertop, or the length of a book. As the year progresses, work with your child to line up an object at a number other than 0 on a ruler and count the intervals to find the length.
- Ask your child to measure two objects and determine how much longer one object is than the other.
- Ask your child to count money. Start with dimes and pennies, and as the year progresses, add nickels, quarters, and dollar bills. Have him or her add the cost of two or three items and determine the amount of change he or she would get back if paying for the items with $1, $5, $10, or $100.
- Ask your child to tell and write time to the nearest hour using an analog clock and using A.M. or P.M. Later in the year, have your child tell and write time to the nearest five minutes.

GEOMETRY

Expectations

Early in the year, your child will be able to recognize 2-dimensional shapes with specific attributes, such as a given number of sides.

As the year progresses, children begin to name 2-dimensional shapes such as triangles, quadrilaterals, pentagons, and hexagons. Your child will describe 3-dimensional shapes and name a cube. He or she will use same-sized square tiles to partition a rectangle into rows and columns, and count to find the total number of tiles. He or she will create larger shapes using triangles and rectangles, and will begin to sort shapes by their attributes.

Later in the year, in preparation for work with area in grade 3, your child will learn to partition rectangles into rows and columns of same-sized squares and count to find the total number of squares.

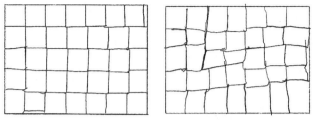

Examples of correct partitioning. © McGraw-Hill Education.

By the end of grade 2, your child can be expected to recognize and draw shapes having specific attributes, such as a given number of angles or a given number of equal faces. He or she will identify triangles, quadrilaterals, pentagons, hexagons, and cubes. Your child will also partition a rectangle into rows and columns of same-sized squares and count to find the total number of squares.

Do-Anytime Activities

• Look for geometric shapes in everyday objects around the house, at the market, in architectural features, and on street signs. Name the shapes using their geometric names and have your child share attributes of the shapes.
• Ask your child to partition rectangles into a specified number of same-sized squares.

ALGEBRA

Expectations

Early in the year, expect your child to determine whether a number less than 20 is even or odd by pairing pennies or counters.

11 is an odd number because one penny does not have a pair.

Your child will apply the Commutative and Associative Properties of Addition to solve problems. For example:

Commutative Property of Addition: If I know 3 + 6 = 9, then I
also know 6 + 3 = 9.
Associative Property of Addition: (2 + 5) + 4 = 2 + (5 + 4),
because 7 + 4 = 2 + 9.

And he or she will use or find rules relating two numbers using
"What's My Rule?" tables. (See examples of "What's My Rule?"
tables in "Algebra in Kindergarten?")

As the year progresses, your child will represent even numbers
as the sum of two equal addends. For example, 10 can be represented by 5 + 5. He or she will also
organize information from a number
story in diagrams and represent the
problem with number models using a
? for the unknown value. For example
*Jessica took out 8 books from the li-
brary. She did not read 3 of the books.
How many books did Jessica read?*

Total	
8	
Part	Part
3	?

3 + ? = 8

By the end of grade 2, expect your
child to use drawings and equations
to represent problems and number
stories involving addition and subtraction; find the unknown
number in addition and subtraction equations, such as 8 + ? = 10,
4 = 8 − ?, 3 + 3 = ?; and understand properties of addition and the
relationship between addition and subtraction. Expect your child
to be able to decide whether a number up to 20 is odd or even and
to write an equation using two equal addends to represent an even
number. For example, 18 is even because 9 + 9 = 18.

Do-Anytime Activities

- Name a set of numbers that change following a rule, and ask
  your child to name the rule. For example, for the numbers 3,
  6, 9, 12, the rule might be *add 3*.
- When practicing number stories, have your child represent
  the number stories by drawing pictures. Later in the year,

have your child represent the number stories with number models using a ? for the unknown number.

- Have your child tell number stories that fit different equations, such as 7 + ? = 12 and ? − 9 = 8.
- Have your child explain why 3 + 2 gives the same answer as 2 + 3.
- Name a number less than 20, and have your child tell whether it is an odd or even number. If it is an even number, have him or her write a number model for the number using two equal addends.

PLACE VALUE WITH WHOLE NUMBERS

Expectations

Early in the year, children begin using open number lines and place value to round whole numbers to the nearest 10 or 100. Your child will learn to use the rounded numbers to make quick, often mental, calculations to estimate, and will use the estimates to check whether exact answers make sense.

Your child will also use close-but-easier numbers to estimate sums and differences. For example, when estimating the difference between 354 and 287, your child might think, "354 is close to 355 and 287 is close to 300, so the difference should be a little more than 55 because I went up to 300 from 287."

By midyear, your child will be expected to use place-value understanding to round whole numbers to the nearest 10 or 100.

By the end of grade 3, your child will be expected to estimate sums and differences using rounding or close-but-easier numbers. He or she will use these estimates to assess the reasonableness of answers. (See also the expectations under third-grade "Operations with Whole Numbers.")

Do-Anytime Activities

- Name a 2-digit number and ask your child which multiple of 10 is closer to that number. For example: *Is 23 closer to 20 or 30? Is 49 closer to 40 or 50?* In the same manner, name a 3-digit number, and ask which multiple of 100 is closer.
- Name a 2- or 3-digit number, and ask your child to name the closest multiples of 10 or 100. For example, 50 and 60 are the closest multiples of 10 to 53; and 200 and 300 are the closest multiples of 100 to 289.
- Have your child estimate sums and differences of 2- or 3-digit numbers and then find the exact answers. Then have your child compare the estimates with the exact answers and judge whether his or her answers and estimates are reasonable.

OPERATIONS WITH WHOLE NUMBERS

Expectations

■ Addition and Subtraction. Early in the year, expect your child to add and subtract within 1,000 to solve problems and number stories using tools such as a number grid and strategies based on place value and the relationship between addition and subtraction. Soon after, your child will review addition and subtraction strategies from grade 2 and will be introduced to column addition, an algorithm that emphasizes place-value understanding.

327 + 198 = ?

	3	2	7
+	1	9	8

Add the numbers in each column.   4   11   15

Trade 10 ones for 1 ten.
Move 1 ten to the tens column.   4   12   5

Trade 10 tens for 1 hundred.
Move the 1 hundred to the hundreds column.   5   2   5

327 + 198 = 525
Column addition

After midyear, your child will be introduced to trade-first subtraction, another algorithm that emphasizes place-value understanding.

431 − 259 = ?

100s	10s	1s
4	3	1
−2	5	9

Look at the 100s place.
Since 400 > 200, there is no trade to make.

100s	10s	1s
3	13	
−4	3	1
−2	5	9

Look at the 10s place.
Since 50 > 30, you need to make a trade with the column to the left.

100s	10s	1s
	12	
3	13	11
−4	3	1
−2	6	9
1	**7**	**2**

Look at the 1s place.
Since 9 > 1, you need to make a trade with the column to the left. Now subtract in each column in any order.

431 − 259 = 172
Trade-first subtraction

By the end of grade 3, expect your child to fluently add and subtract within 1,000 to solve problems and number stories using strategies and algorithms the children have learned.

■ Multiplication and Division. Early in the year, expect your child to use drawings to solve simple multiplication and division number stories. (See also the expectations under third-grade "Algebra.")

• • • •   There are 12 toy cars
• • • •   in all.
• • • •

Multiplication example: Justin has 3 boxes of toy cars. Each box has 4 toy cars. How many toy cars does Justin have?

O O O O O O O O
1 2 1 2 1 2 1 2

Each child gets 4 strawberries.

Division example: Ayanna wants to equally share 8 strawberries with one friend. How many strawberries will each child get?

Midway through the year, your child will solve 2-step number stories involving multiplication and division. (See also the expectations under third-grade "Algebra.")

Late in the year, your child will apply his or her understanding of multiplication as equal groups and place value to develop strategies for multiplying multiples of 10. For example, to solve 5 × 70, your child could use base-10 blocks to model 70 as 7 longs and rename 70 as 7 groups of 10. So 5 × 70 means 5 groups with 7 tens in each group. This totals 35 tens, or 350.

By the end of grade 3, expect your child to solve 1-step number stories involving multiplication and division and 2-step number stories involving any two operations. Also expect your child to multiply 1-digit whole numbers by multiples of 10 less than 100 using strategies based on place value and properties of operations. (See also the expectations under third-grade "Place Value with Whole Numbers.")

Do-Anytime Activities

- With your child, tell number stories about equal groups and equal shares in everyday life. For example: *5 bags with 3 crackers in each bag totals 15 crackers. 12 books shared among 4 children gives each child 3 books.*
- Roll dice or draw number cards to generate two 2- or 3-digit numbers. Then have your child add or subtract the numbers using any strategy.
- Later in the year, tell 2-step number stories, and have your child solve them and explain his or her solution strategies. For example: *Jenny shared 20 pennies with her brother. Then she found 8 more pennies under the bed. How many pennies does she have now? 20 ÷ 2 = 10; 10 + 8 = 18. Jenny now has 18 pennies.*

MATH FACTS

Expectations

Early in the year, children develop strategies for solving 2s, 5s, and 10s multiplication facts. These facts will serve as "helper facts" for more sophisticated strategies. Your child will also be introduced to multiplication/division Fact Triangles and fact families that relate the same sets of 3 numbers.

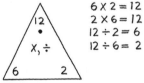

$6 \times 2 = 12$ 2, 6, 12 Fact Triangle and related fact family
$2 \times 6 = 12$
$12 \div 2 = 6$
$12 \div 6 = 2$

Your child will begin to develop an understanding of division as an unknown factor problem. For example, to solve 12 ÷ 6, one can ask 6 times what is 12 (6 × ___ = 12) and get 2. Soon after this, your child will be introduced to the adding-a-group and subtracting-a-group strategies for deriving new facts from familiar multiplication facts, such as multiplication squares (2 × 2, 4 × 4, and so on) and 2s, 5s, and 10s facts.

Adding a Group

```
X X X
X X X
X X X
X X X
X X X
X X X
```

$5 \times 3 = 15$, $15 + 3 = 18$, so $6 \times 3 = 18$

Subtracting a Group

$5 \times 3 = 15$
```
X X X
X X X
X X X
X X X
X X X
```

↓

$15 - 3 = 12$
```
X X X
X X X
X X X
X X X
X X X
```
$4 \times 3 = 12$

Adding-a-group and subtracting-a-group strategies

Midway through the year, your child will be introduced to other fact strategies. These strategies include doubling (when one factor in a multiplication fact is doubled, the product is doubled) and breaking apart (when one factor of an unknown multiplication fact is decomposed to create two easier helper facts).

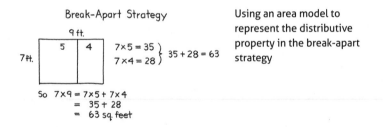

Break-Apart Strategy

9 ft.

| 5 | 4 |

7 ft.

$7 \times 5 = 35$
$7 \times 4 = 28$ } $35 + 28 = 63$

So $7 \times 9 = 7 \times 5 + 7 \times 4$
= $35 + 28$
= 63 sq. feet

Using an area model to represent the distributive property in the break-apart strategy

By the end of grade 3, your child will be expected to fluently multiply and divide within 100 using strategies such as the relationship between multiplication and division. Your child will also be expected to know from memory all products of two 1-digit numbers and to be able to find the unknown whole number in a multiplication or division equation relating three whole numbers.

Do-Anytime Activities

- Make a set of four index cards, and write 1, 2, 5, and 10—one number on each card. Roll a die, and draw a card to generate two numbers. Have your child multiply the two numbers. As the year progresses, add 3, 4, 6, 7, 8, and 9 to the set of index cards.
- Play fact-related games found in the *Student Reference Book* or online in the Student Learning Center.
- Later in the year, pose multiplication problems involving 1-digit numbers times multiples of 10, and ask your child to solve. For example: 7 × 40 = 280.
- Later in the year, pose multiplication problems involving 1-digit numbers multiplied by 2-digit numbers with products less than 100, and ask your child to break apart the 2-digit factor into two easier-to-multiply factors and solve. For example, in 3 × 17, 17 can be broken into 10 and 7. Since 3 × 10 = 30, 3 × 7 = 21, and 30 + 21 = 51, 3 × 17 = 51.

FRACTIONS

Expectations

Early in the year, children learn to divide wholes into equal shares and name equal shares with fractions. At this time, expect your child to write fraction names using word or number-word notation (e.g., *one-half, 1-fourth*). Your child will also be introduced to fraction circle pieces, one of the tools used to continue to develop an understanding of fractions as numbers.

In the middle of the year, children are formally introduced to standard fraction notation ($\frac{1}{2}$, $\frac{2}{3}$, $\frac{3}{4}$, and so on). Your child will learn that the denominator (the number on the bottom of the fraction) represents the number of equal parts needed to make a whole, and the numerator (the number on the top of the fraction) represents the number of parts being considered. At this time, you can expect your child to identify and represent unit fractions, such as $\frac{1}{2}$ and $\frac{1}{4}$, and non-unit fractions, such as $\frac{2}{6}$ and $\frac{3}{4}$, using pictures, words, and fraction circles.

Number of Equal Parts in the Whole	Words for Size of the Parts	Words for One Part of the Whole	Fraction Circles	Number for One Part of the Whole
2	half, halves	1-half, one-half one out of two equal parts		$\frac{1}{2}$
3	third, thirds	1-third, one-third one out of three equal parts		$\frac{1}{3}$
4	fourth, fourths	1-fourth one-fourth, one out of four equal parts		$\frac{1}{4}$

Later in the year, your child will use fraction strips, fraction number lines, and fraction circles to help locate fractions on number lines, identify equivalent fractions, and compare fractions.

By the end of grade 3, your child will be expected to understand a fraction as a number on a number line and represent fractions on number lines. He or she will also be expected to understand equivalent fractions and compare fractions by reasoning about their size.

Do-Anytime Activities

- When serving different foods, discuss ways to cut the food to feed 2, 3, 4, 6, or 8 people so that each person gets an equal share. Ask your child to provide various names for each share and for the whole.
- Ask your child to find examples of fractions around your home and explain what each fraction means. For example: *This $\frac{2}{3}$-measuring cup means 2 of 3 equal parts that make a whole cup. One more third will make 1 whole cup.*
- Later in the year, ask your child to draw a number line from 0 to 1, divide it into 2 equal parts with tick marks, and label

each tick mark with a fraction ($\frac{0}{2}, \frac{1}{2}, \frac{2}{2}$). Do the same with 3, 4, 6, and 8 equal parts.

DATA

Expectations

Early in the year, your child will represent a data set on a scaled bar graph and use the information in the graph to solve "How many more?" and "How many fewer?" problems.

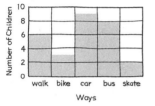

How Bay School 3rd Graders Get to School

A scaled bar graph. How many more children ride to school in a car than on a bike?

Your child will also be introduced to scaled picture graphs, which are graphs constructed with icons that each represent the same number of multiple data points.

Number of Books Checked Out

Key: ☐ = 2 books

A scaled picture graph

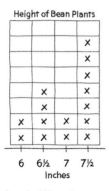

Height of Bean Plants

A scaled line plot

Soon after, your child will be expected to measure lengths to the nearest $\frac{1}{2}$ inch and represent the data on a line plot with the scale marked in whole numbers and halves.

By the end of grade 3, your child can be expected to draw a scaled picture graph and a scaled bar graph to represent a data set with several categories. He or she will be expected to solve 1- and 2-step "How many more?" and "How many less?" problems using information in the scaled bar graphs.

Your child will also be expected to show measurement data on line plots with scales marked with wholes, halves, or fourths. (See also the expectations under third-grade "Measurement.")

Do-Anytime Activities

- Have your child collect data by posing questions to family members and friends and then organize the data in scaled bar graphs or picture graphs. For example: *What is your favorite sport—baseball, basketball, soccer, dance, swimming, or running? Which type of book do you like best—mystery, adventure, science, or sports?* Ask your child "How many more?" and "How many fewer?" questions about the data.
- Have your child measure lengths of items to the nearest $\frac{1}{2}$ inch, such as shoes or shirtsleeves, and plot the data on a scaled line plot. As the year progresses, measure items to the nearest $\frac{1}{4}$ inch.

MEASUREMENT

Expectations

■ Time. Early in the year, expect your child to tell time to the nearest 5 minutes. Soon after, he or she will be introduced to telling time to the nearest minute. Your child will also learn to calculate elapsed time using a toolkit clock or an open number line.

4 + 10 + 15 = 29 minutes.
There are 29 minutes
between 8:46 and 9:15.

By the end of grade 3, expect your child to tell time to the nearest minute and solve number stories involving elapsed time.

■ Length. Early in the year, expect your child to use a ruler to measure lengths to the nearest whole inch.

Midway through the year, your child will be introduced to measuring lengths to the nearest $\frac{1}{2}$ inch and later to the nearest $\frac{1}{4}$ inch.

By the end of grade 3, expect your child to use rulers to measure lengths to the nearest $\frac{1}{4}$ inch and plot the measures on a line plot. (See also the expectations under third-grade "Data.")

■ Mass and Liquid Volume. In the beginning of the year, your child will be introduced to measuring mass in grams and kilograms using a pan balance and standard masses.

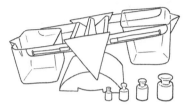
A pan balance and standard masses

A little later, he or she will start measuring liquid volume in liters.

A liter beaker

As the year progresses, third graders are introduced to benchmarks for masses and liquid volume. Your child will explore and label objects with benchmark masses, such as 1 gram, 10 grams, 100 grams, 1 kilogram, and so on. Then he or she will compare the benchmark masses with items with unknown masses and use the benchmarks to estimate the masses. For example, holding an object like a pencil in one hand and a paper clip in the other, your child might think: *I know the paper clip is about 1 gram, so this pencil has more mass than the paper clip.* Then he or she will use a pan balance and standard masses to check the estimates.

Children will explore and learn about liquid volume in a similar manner.

By the end of grade 3, expect your child to estimate and measure liquid volumes and masses of objects using grams, kilograms, and liters. He or she will also be able to add, subtract, multiply, and divide to solve 1-step word problems involving masses or volumes, using drawings to represent the problem.

■ Area and Perimeter. Midway through the year, your child will be introduced to perimeter and area. At this time, expect your child to distinguish between area and perimeter, to measure areas by tiling and counting squares, and to find the perimeters of shapes with given side lengths.

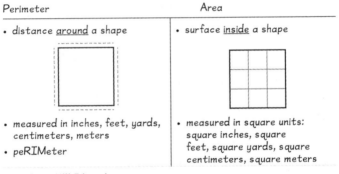

Perimeter	Area
• distance _around_ a shape	• surface _inside_ a shape
• measured in inches, feet, yards, centimeters, meters • peRIMeter	• measured in square units: square inches, square feet, square yards, square centimeters, square meters

© McGraw-Hill Education.

By the end of grade 3, expect your child to be able to find the areas of rectangles by multiplying side lengths. He or she will also be able to find the perimeters of polygons, find missing side lengths, and draw rectangles with the same perimeter and different areas or with the same area and different perimeters.

Do-Anytime Activities

- With your child, find containers in your home that hold about 1 liter and $\frac{1}{2}$ liter of liquid to use as liquid volume benchmarks. Your child can pour water into various other containers to estimate their volumes.

- A nickel has about 5 grams of mass. Have your child hold a nickel in one hand and a different small item in the other and compare the masses of the two items. Ask: *Do they have the same mass? Which is more? Less?* In a similar manner you can use 20 nickels (about 100 grams) or 100 nickels (about 500 grams or $\frac{1}{2}$ kilogram).
- Ask your child to tell time to the nearest 5 minutes using an analog clock and explain how he or she knew the hour. As the year progresses, have your child tell and write time to the nearest minute.
- Have your child measure lengths of items in and around your home, including toys, plants, and so on, to the nearest $\frac{1}{4}$ inch.
- You and your child can measure the side lengths of rectangular rooms in your home; then have your child find the area and perimeter of each.

GEOMETRY

Expectations

Early in the year, expect your child to recognize and draw polygons (2-dimensional, closed figures with straight lines that do not intersect) having specific attributes, such as a given number of angles or a given number of sides.

Midway through the year, expect your child to understand that polygons that are in different categories may nevertheless share attributes. For example, a rectangle and a square both have four right angles and two pairs of parallel sides, so a square is also a rectangle. However, since a square has four sides of equal length, a rectangle is not always a square.

By the end of grade 3, expect your child to understand that different polygons—rectangles and rhombuses, for example—may share attributes and that the shared attributes can define a larger category of shapes, such as quadrilaterals (four-sided polygons). Your child should be able to recognize rhombuses, squares, and rectangles as examples of quadrilaterals, and draw examples of quadrilaterals that do not belong to any of these subcategories of shapes.

This quadrilateral is not a square, rectangle, rhombus, kite, trapezoid, or parallelogram.

Do-Anytime Activities

- Look for quadrilaterals in everyday objects around the house, at the market, in architectural features, and on street signs, and have your child describe the attributes.
- Have your child draw quadrilaterals that fit the following descriptions: all sides equal in length, at least one right angle, a pair of parallel sides, no sides equal in length, two pairs of parallel sides, and no right angles.

A trapezoid is a quadrilateral with at least 1 pair of parallel sides.

ALGEBRA

Expectations

Early in the year, expect your child to represent simple multiplication and division number stories with drawings and solve them.

As the year progresses, your child will use diagrams to help organize information from multiplication and division 1- and 2-step number stories. Your child will use symbols (such as question marks) or letters to represent the unknown quantity in his or her number models.

number of pages	stamps per page	stamps in all
P	9	36

P is the number of pages.

$P \times 9 = 36$ or $36 \div 9 = P$

You need 4 pages.

1-step number story: *You have 36 stamps. You want to put 9 stamps on each page. How many pages do you need?*

Change
Start		End
4 × 8	−6	M

M is the number of markers left.

$4 \times 8 = 32$

$32 - 6 = M$

Gregory has 26 markers left.

2-step number story: *Gregory has 4 packs of markers. Each pack has 8 markers. He gives 6 markers to his friend. How many markers does Gregory have left?*

Your child will apply the Commutative, Associative, and Distributive Properties to solve problems. Note that your child will not be expected to know the formal names for the properties.

Commutative Property of Multiplication: If I know $3 \times 6 = 18$, then I also know $6 \times 3 = 18$.

Associative Property of Multiplication: $2 \times 5 \times 4$ can be found by $2 \times 5 = 10$, then $10 \times 4 = 40$; or by $2 \times 4 = 8$, then $8 \times 5 = 40$.

Distributive Property: Knowing that $6 \times 5 = 30$ and $6 \times 2 = 12$, one can find 6×7 as $6 \times (5 + 2) = (6 \times 5) + (6 \times 2) = 30 + 12 = 42$.

Expect your child to be able to determine the unknown whole number in a multiplication or division equation relating three whole numbers. For example: $12 \div ? = 3$ and $24 \div ? = 6$. Your child will also be introduced to order of operations, which indicates the order in which calculations are carried out in number sentences:

ORDER OF OPERATIONS*

1. Do operations inside parentheses first. Follow rules 2 and 3 when computing inside parentheses.
2. Multiply and divide in order, from left to right.
3. Add and subtract in order, from left to right.

---

* Note that this differs from the order of operations shown in the algebra expectations for sixth grade because third graders have not yet begun work with exponents or roots.

By the end of grade 3, children are expected to use multiplication and division to solve 1- and 2-step number stories and to use drawings and equations with a symbol or letter for the unknown quantity to represent the problems and number stories. Expect your child to apply properties of operations to multiply and divide and determine the unknown in multiplication and division equations.

Do-Anytime Activities

- Pose 1 and 2 step number stories for your child to solve and represent with drawings and number models. Later in the year, have your child represent the number stories using number models with symbols or letters standing for the unknown quantity.
- Have your child tell number stories that fit different equations, such as $4 \times 6 = 24$ and $21 \div 3 = 7$. For example: *I have 4 bags with 6 grapes in each bag, so I have 24 grapes in all.* Or, *I have 21 grapes and want to share them among my brother, my sister, and myself so each of us gets 7 grapes.* Later in the year, pose equations that involve more than one operation like $(8 \div 2) - 2 = 2$. For example: *My sister and I have 8 balloons and share them equally between us. We each have 4 balloons. Then 2 of my balloons pop, so now I only have 2 balloons.*
- Have your child solve problems with unknown factors or products, such as $18 = 6 \times$ ___, $4 \times 9 =$ ___, or ___ $= 8 \times 4$.
- Have your child explain how knowing $4 \times 7 = 28$ helps to know $28 \div 4 = 7$.

FOURTH GRADE

NUMBERS AND COUNTING

Expectations

Early in the year, your child will work on identifying more than one factor pair for a given number and writing multiples of a 1-digit number. For example, the factor pairs for 12 are 1 and 12, 2 and 6, and 3 and 4. Multiples of 3 are 3, 6, 9, 12, 15, 18, 21, and so on. Your child will also identify prime and composite numbers less than 40. A prime number is a counting number (1, 2, 3, 4, and so on) greater than 1 that has only two whole-number factors, 1 and itself. For example, 7 is a prime number because its only factors are 1 and 7. Composite numbers are counting numbers that have more than two factors.

By the middle of the year, your child will be able to find all of the factor pairs for numbers from 1 through 100 and will recognize that a whole number is a multiple of each of its factors. For example, 5 is a factor of 20. Thus, 20 is a multiple of 5. Your child will be expected to determine whether a number up to 100 is a multiple of a given 1-digit number and tell whether a whole number up to 100 is prime or composite. The work fourth graders do with factors and multiples is complete after the middle of the year.

Do-Anytime Activities

- Say a number to your child and have him or her determine whether it is prime or composite. Then ask your child to give you a number with the same question.
- Name a composite number up to 100, and have your child name some, or all, of its factor pairs.

PLACE VALUE WITH WHOLE NUMBERS

Expectations

Early in the year, expect your child to round whole numbers through the hundred thousands place to the two largest places in the number. For example, 28,478 rounded to the nearest ten thousand is 30,000 and rounded to the nearest thousand is 28,000. Your child should be able to read and identify places in whole numbers through the hundred thousands, read number names through the hundred thousands, read numbers in expanded form through the hundred thousands, and write numbers in expanded form through the thousands.

10,000s ten thousands	1,000s thousands		100s hundreds	10s tens	1s ones
7	6		3	4	2

The number 76,342 is shown in the place-value chart above.

The value of the 7 is 70,000 or 7 (10,000s).
The value of the 6 is 6,000 or 6 (1,000).
The value of the 3 is 300 or 3 (100s).
The value of the 4 is 40 or 4 (10s).
The value of the 2 is 2 or 2 (1s).

76,342 is read as "seventy-six thousand, three hundred forty-two."

Expanded form: 76,342 = 70,000 + 6,000 + 300 + 40 + 2

Also early in the year, your child will compare and order multi-digit whole numbers through the hundred thousands and record the comparisons using >, <, or =. For example, 527,487 > 527,123.

By midyear, your child will be expected to recognize that in a multidigit whole number, a digit in one place represents 10 times what it represents in the place to its right. For example, the 7 in 700 is 10 times as large as the 7 in 70.

By the end of grade 4, your child will be expected to read and write multidigit whole numbers using base-10 numerals, number names, and expanded form; compare two multidigit numbers based on meanings of the digits in each place, using >, =, and < symbols to record the results of comparisons; and use place-value understanding to round multidigit whole numbers to any place.

Do-Anytime Activities

- Look for large numbers on signs or in the newspaper. Have your child practice reading the numbers aloud and identifying the value of each digit.
- Have your child look for numbers and round them to various places.

- You and your child each write down a large number. Read the numbers aloud and then compare them, discussing which one is greater and how you know.

OPERATIONS WITH WHOLE NUMBERS

Expectations

Early in the year, your child will use the U.S. traditional algorithm for addition to solve 4-digit + 4-digit problems and the U.S. traditional algorithm for subtraction to solve 4-digit – 4-digit problems. (See examples of U.S. traditional algorithms in "Moving On Up . . . Operations with Bigger Numbers.") He or she will use fact extensions to multiply by a multiple of 10, solving problems like 8 * 80 = 640 or 70 * 60 = 4,200. Your child will be able to make a plan for solving multistep addition and subtraction number stories, solve the number stories, and assess the reasonableness of answers by comparing them to an estimate.

By the middle of grade 4, your child should fluently add and subtract multidigit whole numbers using the U.S. traditional addition and subtraction algorithms. He or she will accurately multiply 2-digit by 1-digit whole numbers. He or she will also make sense of multistep number stories involving multiplication, make a plan for solving these problems, solve the problems, and assess the reasonableness of answers by estimating answers and comparing the estimates to their calculated answers. After midyear your child will begin dividing a 2-digit number by a 1-digit number and then illustrating and explaining the division.

By the end of grade 4, your child will be expected to multiply a whole number of up to four digits by a 1-digit whole number, multiply two 2-digit numbers, and find whole-number quotients and remainders with up to 4-digit dividends and 1-digit divisors. He or she will also be expected to illustrate and explain the calculations.

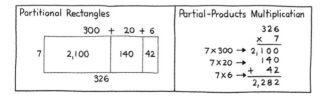

Also expect your child to be able to solve multistep word problems using the four operations, including division problems with remainders.

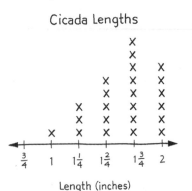

Do-Anytime Activities

- Pose different multistep number stories for you and your child to solve together. Encourage him or her to explain how he or she solved the problem.
- Give your child a multidigit computation problem (addition, subtraction, multiplication, or division). Have your child solve it two different ways and explain the solutions.

DATA

Expectations

Starting after midyear, fourth graders organize and represent data in fractions of a unit (halves and fourths) on line plots. Your child will also solve addition and subtraction problems using the information presented in a line plot.

Cicada Lengths

If you lined up all the cicadas that have a length of $1\frac{1}{4}$ inches, how long would they be?

```
            X
            X
            X    X
        X   X    X
        X   X    X
    X   X   X    X
    X   X   X    X
X   X   X   X    X
+---+---+---+---+---+--->
3/4  1  1 1/4 1 2/4 1 3/4  2
```

Length (inches)

By the end of grade 4, your child will be expected to create a line plot for data in fractions of a unit (halves, fourths, and eighths). He or she will also be expected to solve problems involving addition and subtraction of fractions using information presented in line plots.

Do-Anytime Activities

- Have your child survey several family members and friends to determine the lengths of their pinky fingers to the nearest $\frac{1}{4}$ or $\frac{1}{8}$ inch. Then have him or her plot the data on a line plot. Have your child find the difference between the longest and shortest pinky length.
- Brainstorm other data with your child that could be collected with fractional units. Create a line plot, plot the data, and ask each other questions about the data that can be solved using addition and subtraction.

GEOMETRY

Expectations

Early in the year, expect your child to draw and label points, lines, line segments, and rays. He or she should also correctly identify right angles. Your child will identify properties of line segments and angles within quadrilaterals and identify right angles within triangles.

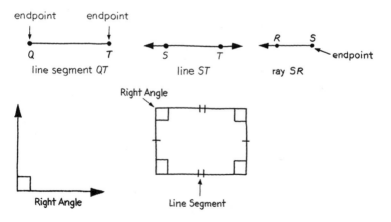

Your child will also be able identify at least one line of symmetry in 2-dimensional symmetric figures.

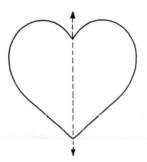

The dashed line is a line of symmetry because it divides the heart into two parts that are mirror images of each other.

As the year progresses, your child will identify lines, line segments, and rays alone or within figures. Your child will draw and represent right angles, identify other angles as acute or obtuse, and draw, represent, and identify perpendicular and parallel lines.

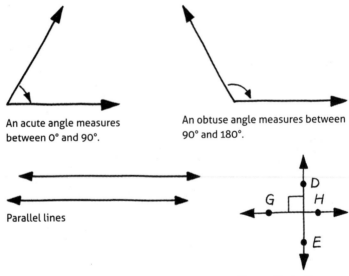

An acute angle measures between 0° and 90°.

An obtuse angle measures between 90° and 180°.

Parallel lines

Perpendicular lines

By midyear, expect your child to be able to classify 2-dimensional figures based on whether they have parallel or perpendicular lines and whether they have angles of a specified size. He or she should be able to recognize right triangles as a category of triangles and identify right triangles.

By the end of grade 4, your child should be able to identify a line of symmetry for a 2-dimensional figure, identify line-symmetric figures, and draw lines of symmetry.

Do-Anytime Activities

- Have your child identify real-word examples of symmetric objects, pointing out the line(s) of symmetry.
- Cut out symmetric shapes from magazines and newspapers. Have your child fold the shape along the line(s) of symmetry and identify the matching parts.
- Together with your child, identify real-world examples of right, acute, and obtuse angles, line segments, parallel lines, and perpendicular lines.

MEASUREMENT

Expectations

Early in the year, your child will use strategies to find the perimeter (the distance around a 2-dimensional figure) and the area (the amount of surface inside a 2-dimensional shape) of various shapes. He or she will then begin to use formulas to find the area and perimeter of rectangles.

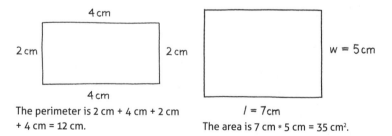

The perimeter is 2 cm + 4 cm + 2 cm + 4 cm = 12 cm.

l = 7 cm

The area is 7 cm * 5 cm = 35 cm².

Your child will also convert among units of time, length, capacity, weight, and mass using a 2-column table and explain the relationship between the columns. He or she will solve number stories involving measurement units. For example, *it takes Jerry 7 minutes to walk to school. It takes Kania 12 minutes. How many more seconds does it take Kania?*

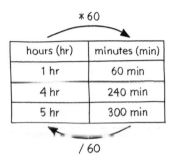

hours (hr)	minutes (min)
1 hr	60 min
4 hr	240 min
5 hr	300 min

Beginning in the middle of the year, children identify rotations (or turns) as $\frac{1}{4}, \frac{1}{2}, \frac{3}{4}$, or full turns. They understand that a measure of 1 degree is an angle that is $\frac{1}{360}$ of a circle and that angles are measured with repetitions of 1-degree angles.

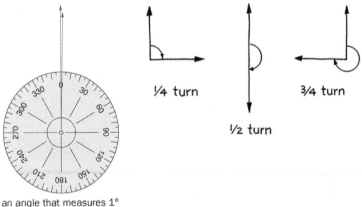

an angle that measures 1°
© McGraw-Hill Education.

Children also learn how to estimate the measure of an angle and then measure it to check the estimate. When given one ray, your child will sketch an angle to a specified number of degrees. He or she will also add and subtract to find unknown angle measures within angles measuring 90 and 180 degrees.

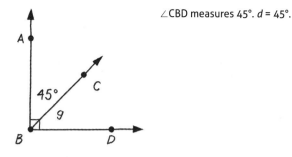

∠CBD measures 45°. *d* = 45°.

By the end of grade 4, your child will use the four operations to solve word problems involving distances, intervals of time, liquid volumes, masses of objects, and money, including problems involving simple fractions or decimals.

Do-Anytime Activities

- Play "Simon Says" using rotations. For example: *Simon says rotate your body ¼ turn.*
- Have your child identify different angles in real-life objects and estimate their measures.
- Look for opportunities to convert measurements in the grocery store or at home.

FRACTIONS

Expectations

Early in the year, your child will use models such as fraction circles and fraction strips to explain why two fractions are equivalent. He

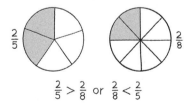

or she will compare and order fractions, using >, =, or < to record the comparisons and will use a model to justify the comparisons. Your child will decompose (or break down) fractions and represent them with an equation. For example, $\frac{4}{8} = \frac{1}{8} + \frac{1}{8} + \frac{1}{8} + \frac{1}{8}$. He or she will explain the equation using models such as fraction circles.

As the year progresses, children will be expected to recognize and generate equivalent fractions ($\frac{1}{2}, \frac{2}{4}, \frac{5}{10}, \frac{25}{50}$, and so on) and explain why two fractions are equivalent. Your child will compare two fractions with different numerators and different denominators by rewriting them with common denominators or numerators. For example: Which is greater, $\frac{3}{6}$ or $\frac{6}{10}$? $\frac{3}{6} = \frac{6}{12}$ and $\frac{6}{12}$ is smaller than $\frac{6}{10}$ (since twelfths are smaller than tenths), so $\frac{3}{6} < \frac{6}{10}$. Children will also be expected to decompose a fraction in more than one way. For example, your child might write $\frac{4}{8} = \frac{1}{8} + \frac{1}{8} + \frac{1}{8} + \frac{1}{8}$ and also $\frac{4}{8} = \frac{2}{8} + \frac{2}{8}$.

In the middle of the year, fourth graders begin adding and subtracting mixed numbers ($3\frac{1}{2}$, $5\frac{3}{8}$, and so on) with like denominators using fraction circle pieces, drawings, and number lines. Your child will also add and subtract fractions in number stories using fraction circle pieces, drawings, and number lines. He or she will use repeated addition and then multiplication to multiply a fraction by a whole number. For example, $3 * \frac{1}{2} = \frac{1}{2} + \frac{1}{2} + \frac{1}{2} = \frac{3}{2} = 1\frac{1}{2}$.

By the end of grade 4, your child will be expected to add and subtraction fractions and mixed numbers with like denominators without using manipulatives; add and subtract fractions in number stories; solve word problems involving multiplication of a fraction by a whole number using fraction circle pieces, drawings, and equations; and add two fractions with denominators of 10 and 100. For example: $\frac{2}{10} + \frac{40}{100} = $ ___. $\frac{2}{10} = \frac{20}{100}$, and $\frac{20}{100} + \frac{40}{100} = \frac{60}{100}$, so $\frac{2}{10} + \frac{40}{100} = \frac{60}{100}$.

Do-Anytime Activities

- Discuss fractions as you are cooking. Have your child find equivalent fractions for given measures.
- Double or triple a recipe with your child to explore multiplying a fraction by a whole number.

DECIMALS

Expectations

Early in the year, your child will represent decimals to hundredths base-10 blocks and with base-10 numerals.

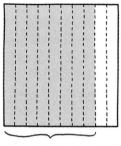

0.8 = eight-tenths = 8/10

Your child will compare and order decimals using a model and will record decimal comparisons. He or she will then justify comparisons of decimals using a model.

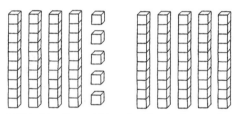

0.45 is 4 tenths and 5 hundredths; 0.5 is 5 tenths.
0.45 is less than 0.5.
0.45 < 0.5

By the end of grade 4, your child will be expected to use decimal notation to represent fractions with denominators of 10 or 100; compare two decimals to the hundredths place; and record the results of comparisons with the symbols >, =, or <.

Do-Anytime Activities

• Together with your child, look for examples of decimals in real-world examples. Practice reading the decimals you find.

ALGEBRA

Expectations

In the beginning of the year, your child will explore comparison number stories and begin to determine which comparison situations are additive and which are multiplicative.

> Additive Comparison: Asha has 5 books. Zach has 3 more books than Asha. How many books does Zach have?
> 5 + 3 = 8; Zach has 8 books.
> Multiplicative Comparison: Noah has $3. Sasha has 4 times as much as Noah. $3 * 4 = $12; Sasha has $12.

Your child will apply addition, subtraction, multiplication, and division rules to "What's My Rule?" tables and will also extend simple shape patterns. (See examples of "What's My Rule?" tables in "Algebra in Kindergarten?")

Later in the year, children will write comparison number stories to fit given equations. For example: the equation $3 * 4 = $12 could represent the number story: *Noah has $3 and Sasha has 4 times as much as Noah.* Your child will also write equations to fit given comparison number stories.

By the end of grade 4, your child will write number models with letters standing for the unknown quantity (e.g., $30 + a = 180$) to fit given addition, subtraction, and multiplication number stories.

Do-Anytime Activities

- Look for real-world examples of numeric and shape patterns.
- Write multiplicative comparison stories with your child. Discuss how to solve them.

PLACE VALUE WITH WHOLE NUMBERS

Expectations

Early in the year, your child will write whole numbers in expanded form and should be able to identify the values of digits in whole numbers.

EXPANDED FORM

$365{,}239 = 3 \,[100{,}000\text{s}] + 6\,[10{,}000\text{s}] + 5\,[1{,}000\text{s}] + 2\,[100\text{s}] +$
$\quad 3\,[10\text{s}] + 9\,[1\text{s}].$

He or she will multiply whole numbers by powers of 10 (10^2, 10^3, 10^6, and so on) and notice patterns in the number of zeroes in the product when doing so.

By midyear, your child should recognize that in any multidigit whole number, a digit in one place represents 10 times as much as it represents in the place to its right and $\frac{1}{10}$ of what it represents in the place to its left.

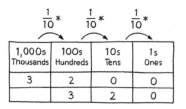

10 *	10 *	10 *	
1,000s Thousands	100s Hundreds	10s Tens	1s Ones
			7
		7	5
	7	5	0
7	5	0	0

A place-value chart showing
the *10 times as much*
relationship between places

$\frac{1}{10}$ *	$\frac{1}{10}$ *	$\frac{1}{10}$ *	
1,000s Thousands	100s Hundreds	10s Tens	1s Ones
3	2	0	0
	3	2	0

A place-value chart showing the
$\frac{1}{10}$ *of* relationship between places

By the end of grade 5, children should understand the relationships between places in multidigit numbers and be able to use them to explain patterns in the number of zeroes in the products when multiplying a whole number by a power of 10.

Do-Anytime Activities

- Look for whole numbers in newspapers, magazines, or other reading materials. Ask your child to read the numbers aloud to you and identify the value of specific digits in the number. For example: *What does the 5 represent in 685,429?*
- Make up "number riddles" for your child to solve like this one: *I'm thinking of a 6-digit number. My number has a 5 in the thousands place, a 2 in the tens place, a 7 in the hundred thousands place, and a 4 in every other place. What number am I thinking of?*
- Ask your child to multiply a whole number by 10, 100, and 1,000. Then have him or her explain what happens to the value of the digits in the number each time he or she multiplies.
- Play *High-Number Toss* with your child. The directions for this game can be found in the grade 5 *Student Reference Book* or online in the Student Learning Center.

NUMBER SYSTEMS

Expectations

Fifth graders begin working with coordinate grids around midyear. Your child will be introduced to the coordinate system, and he or she

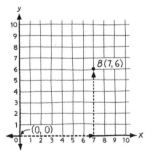

will learn how to name points on a coordinate grid using the location along the horizontal axis (the *x*-axis) first and using the location along the vertical axis (*y*-axis) second. At this point, given some reminders about which coordinate corresponds to which axis, your child should be able to plot and name points on a coordinate grid.

As the year progresses, your child will use coordinate grids to represent mathematical and real-world problems.

By the end of grade 5, children use data from mathematical and real-world situations to form ordered pairs and graph them. Your child will use the graph to answer questions and solve problems.

Fifth-grade *Math Journal* page.
© McGraw-Hill Education.

Do-Anytime Activities

• Give your child two different rules, such as + 1 and * 2, and a starting number of 1 for both rules. Ask your child to write the first five numbers in each pattern and use the patterns to write and graph ordered pairs. Have your child explain what he or she notices about the graph.

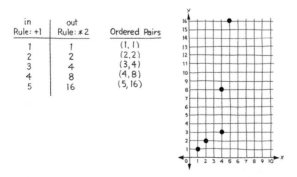

in Rule: +1	out Rule: * 2	Ordered Pairs
1	1	(1, 1)
2	2	(2, 2)
3	4	(3, 4)
4	8	(4, 8)
5	16	(5, 16)

- Play the games *Hidden Treasure* and *Over and Up Squares* with your child. You can find the directions to these games in the grade 5 *Student Reference Book* or online in the Student Learning Center.

OPERATIONS WITH WHOLE NUMBERS

Expectations

Early in the year, your child will be introduced to U.S. traditional multiplication. This method can be tricky at first, so do not expect him or her to use it to solve all multiplication problems. However, he or she should be able to apply it to problems that involve multiplying a 2- or 3-digit number by a 1-digit number. Your child should also be able to use partial-quotients division to solve division problems with a 1-digit divisor. (See examples of U.S. traditional multiplication and partial-quotients division in "Moving On Up . . . Operations with Bigger Numbers.")

By around midyear, children should better understand and be able to apply the steps of U.S. traditional multiplication and be able to compare it to other multiplication methods he or she knows. Your child should also be able to solve division problems using partial-quotients division and understand what to do with remainders in number stories.

IF THE LIBRARY HAS $80 TO BUY BOOKS THAT ARE $17 EACH, HOW MANY BOOKS CAN THE LIBRARY BUY?

The answer to the division problem is 4 remainder 12. Should she ignore the remainder or report it as a fraction?

By the end of grade 5, children should be able to explain the partial-quotients division algorithm and use it to divide 4-digit numbers by 2-digit numbers. Furthermore, your child should be able to use U.S. traditional multiplication to solve multiplication problems, although it is fine if he or she prefers to use a different method to multiply.

Do-Anytime Activities

- Ask your child to solve a multiplication problem using U.S. traditional multiplication and another multiplication strategy. Have your child explain which strategy he or she prefers and why.
- Have your child tell you a division number story that involves a remainder. Ask him or her to solve the number story and explain how he or she knew what to do with the remainder.
- Ask your child to explain the steps of the partial-quotients division algorithm and make a drawing to show why it works.
- Play the game *Division Dash* and *Multiplication Top-It: Larger Numbers* with your child. The directions to both of these games can be found in the grade 5 *Student Reference Book* or online in the Student Learning Center.

DATA

Expectations

Work with data does not begin until after midway through grade 5, when children learn to use fractional measurement data to create a line plot when given a pre-labeled line. Your child should be able to answer simple questions about the data shown in the line plot.

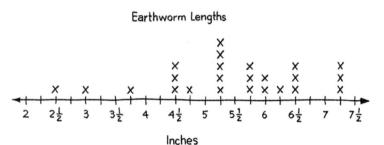

Earthworm Lengths

Inches

A line plot showing the lengths of earthworms found on the playground at recess

By the end of grade 5, your child can be expected to make line plots that show fractional data and solve multistep problems about the data.

Do-Anytime Activities

- Along with your child, brainstorm measurement data that can be collected in fractional units, such as the height of different plants or the number of minutes it takes to run down the block and back. Ask your child to collect the measurements, use them to create a line plot, and pose questions about the line plot.
- Look for examples of line plots in newspapers, magazines, or online. Ask your child questions about the data shown in the line plots.

GEOMETRY

Expectations

Fifth graders do not begin work with geometry until late in the year. At this point, your child will be expected to use attributes of 2-dimensional shapes to define categories and subcategories. He or she will be introduced to hierarchies, which are systems of classification that show the relationships between categories and subcategories.

Triangles   Triangle hierarchy

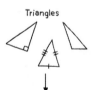

Isosceles triangles

Equalateral triangles

By the end of grade 5, children will be able to classify 2-dimensional figures in a hierarchy. Your child should understand how a hierarchy shows the relationships between categories and subcategories.

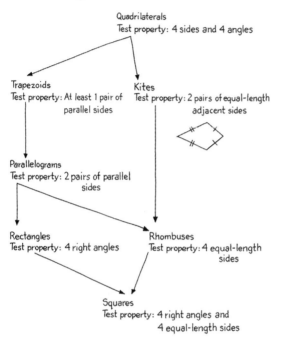

Quadrilaterals
Test property: 4 sides and 4 angles

Trapezoids
Test property: At least 1 pair of parallel sides

Kites
Test property: 2 pairs of equal-length adjacent sides

Parallelograms
Test property: 2 pairs of parallel sides

Rectangles
Test property: 4 right angles

Rhombuses
Test property: 4 equal-length sides

Squares
Test property: 4 right angles and 4 equal-length sides

A quadrilateral hierarchy showing where to classify Shape B. Shape B has 4 sides and 4 angles, so it is a quadrilateral. It has 2 pairs of equal-length adjacent sides, so it is a kite. The 4 sides are not equal length, so it is not a rhombus. A kite is the most specific name for Shape B.

Do-Anytime Activities

- Play the game *Property Pandemonium*. The directions to this game can be found in the grade 5 *Student Reference Book* or online in the Student Learning Center.
- Pay attention to and discuss things in your child's world that can be classified into categories and subcategories. For example: *Our pet is an animal. Our pet is also a dog, which is a subcategory of animals. Our pet is*

Animal

Dog

Golden Retriever

also a golden retriever, which is a subcategory of both animals and dogs.

- Ask your child to create a hierarchy illustrating the relationships between the categories and subcategories described in the previous bullet.

MEASUREMENT

Expectations

Early in the year, expect your child to make simple measurement conversions, such as changing inches to feet or centimeters to meters. Your child will also learn about volume. He or she should be able to describe how volume measures the amount of space an object takes up and understand that it makes sense to measure volume with cubes. He or she will fill rectangular prisms (boxes) with cubes and see that the number of cubes it takes to fill a prism is the same as the volume of the prism.

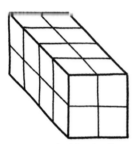

This prism holds 16 cubes, so its volume is 16 cubic units.

As the year progresses, your child will learn that he or she can find the volume of a rectangular prism using two different formulas. He or she can multiply the length, width, and height of a prism ($V = l \times w \times h$) or multiply the area of the base of a prism by its height to calculate volume ($V = B \times h$). (The latter method works for all prisms, not only rectangular prisms.)

Volume $= 3$ feet $* 6$ feet $* 2$ feet $= 36$ feet^3 Volume $= 4$ inches^2 $* 6$ inches $= 24$ inches^3

By the end of grade 5, expect your child to use measurement conversions to solve multistep, real-world problems. He or she should also be able to choose and apply formulas for volume, and also find the volume of shapes composed of more than one rectangular prism by finding the volume of each smaller prism and adding the volumes together.

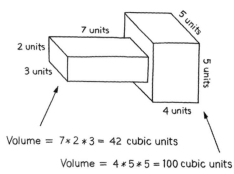

Volume = 7 * 2 * 3 = 42 cubic units

Volume = 4 * 5 * 5 = 100 cubic units

Volume = 42 + 100 = 142 cubic units

Do-Anytime Activities

- Ask your child to measure an object in a given unit. For example: *What is the perimeter of your closet in inches?* Then ask him or her to report the same measurement in a different unit. For example: *What is the perimeter of your closet in feet? In yards?*
- Show your child two different boxes. Ask him or her to predict which box has a greater volume. Then have him or her calculate the volume of each box to see whether the prediction was correct.
- Look for real-world objects that can be modeled by rectangular prisms. Ask your child to estimate the volume of the objects. For example: *This chair could be modeled by six rectangular prisms. What is the approximate volume of the chair?*

Modeling a chair with rectangular prisms

FRACTIONS

Expectations

Early in the year, your child will use drawings and fraction circle pieces to solve problems like this one: *Niko split a pizza evenly with 2 of his friends. How much pizza did each child get?*

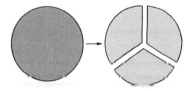

Using fraction circle pieces to show 1 pizza split evenly among 3 friends means each person will get $\frac{1}{3}$ pizza

Children now begin to understand the connection between division and fractions. For example, your child will understand that a pizza divided among 3 friends will be $\frac{1}{3}$ pizza for each. He or she will also use tools like fraction circle pieces, drawings, and number lines to solve fraction and mixed-number addition and subtraction problems.

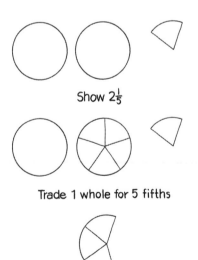

Using fraction circle pieces to solve $2\frac{1}{5} - 1\frac{3}{5}$

Show $2\frac{1}{5}$

Trade 1 whole for 5 fifths

Take away $1\frac{3}{5}$. $\frac{3}{5}$ is left.

Early in the year, fifth graders also solve "fraction-of" problems using counters, drawings, or division. For example, he or she might draw a picture like the one below to show that $\frac{1}{4}$ of 48 is 12

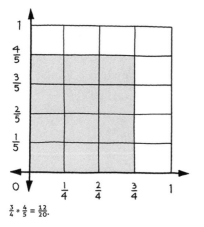

Around midyear, your child will learn strategies for finding common denominators and use these strategies to solve addition and subtraction problems involving fractions and mixed numbers with unlike denominators. He or she will also explore different methods of solving fraction multiplication problems and will find that he or she can multiply fractions by multiplying the numerators of the fractions to find the numerator of the product and multipling the denominators of the fractions to find the denominator of the product.

$\frac{3}{4} * \frac{4}{5} = \frac{12}{20}$.

Your child will also use drawings to help him or her divide whole numbers by unit fractions (fractions with 1 in the numerator) and unit fractions by whole numbers.

Four family members share $\frac{1}{4}$ of a pan of corn bread.
How much of the pan will each person get?

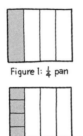

Figure 1: $\frac{1}{4}$ pan

Figure 2: $\frac{1}{4}$ pan ÷ 4 people

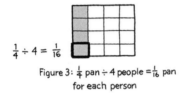

$\frac{1}{4} \div 4 = \frac{1}{16}$

Figure 3: $\frac{1}{4}$ pan ÷ 4 people $= \frac{1}{16}$ pan
for each person

By the end of grade 5, children are expected to solve a variety of fraction computation problems, as well as number stories that involve computation with fractions. Your child should be able to add and subtract fractions and mixed numbers with unlike denominators; multiply fractions by whole numbers, fractions, and mixed numbers; multiply mixed numbers by whole numbers and mixed numbers; and divide unit fractions by whole numbers and whole numbers by unit fractions.

Do-Anytime Activities

- Ask your child to tell number stories that involve sharing or splitting things equally. Ask him or her to solve these problems and use a drawing or a number model to explain the solution.
- Look for real-world situations in which computations are made with fractions. Ask your child to help you in these situations. For example: *The meatloaf recipe calls for 1$\frac{1}{2}$ teaspoon of garlic powder, but I only want to use half that amount. How much garlic powder should I use?*

- Play a fraction game with your child. A wide variety of fraction games can be found in the grade 5 *Student Reference Book* or online in the Student Learning Center.

DECIMALS

Expectations

Work with decimals begins around midyear in grade 5. Your child will now be expected to represent decimals through thousandths on grids and read and write decimals through thousandths, although children at this time are not expected to write decimals with placeholder zeroes, such as in 0.907.

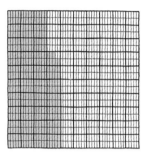

0.368 represented on a thousandths grid

Also around midyear, your child will begin to apply the patterns that he or she knows about place value in whole numbers to decimals. He or she should be able to compare and order decimals with the same number of digits and use tools, such as grids or number lines, to round decimals to the nearest tenth or hundredth. Your child will use grids to add and subtract decimals through hundredths.

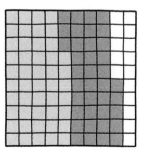

Shading a grid to add 0.47 + 0.38

As the year continues, children use addition and subtraction methods for whole numbers to add and subtract decimals through hundredths. For example, your child might use partial sums to add decimals or U.S. traditional subtraction to subtract decimals. (See examples of partial-sums addition and U.S. traditional subtraction in "Moving On Up . . . Operations with Bigger Numbers.") Your child will also be able to represent, read, and write any decimal through thousandths; explain how place-value patterns apply to decimals; and round decimals through thousandths.

By the end of grade 5, your child should be able to multiply and divide decimals through hundredths.

Do-Anytime Activities

- Look for decimals in newspapers, online, or on objects like packages of food. Ask your child to say the decimals aloud and order the decimals from smallest to largest.
- Tell number stories involving decimals. Ask your child to solve the number stories and explain the method that he or she used.
- With your child play one of the many decimal games, such as *Decimal Domination*, *Decimal Top-It*, and *Doggone Decimal*, found in the *Student Reference Book* or online in the Student Learning Center.

ALGEBRA

Expectations

Early in the year, expect your child to use grouping symbols (like parentheses or brackets) in math expressions and to evaluate expressions with grouping symbols. He or she will also use expressions to model number stories and mathematical problems. For example, your child might write (6 + 3) * 2 when asked: *What is the sum of 6 and 3 multiplied by 2?* As the year continues, your child should be able to model more complex number stories with expressions, including multistep number stories.

Later in the year, your child will begin to extend numerical patterns using given rules. He or she will also identify rules for patterns.

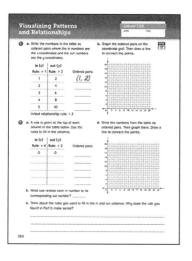

Do-Anytime Activities

- Describe a sequence of operations, such as: *Subtract 12 from 16 and then multiply the difference by the sum of 2 and 4.* Have your child use numbers and symbols, including parentheses and brackets when needed, to write what you describe in words.

- When telling and solving number stories, have your child write a number model using a letter for the unknown and then solve the number story.

- Give your child a rule, such as $+ 3\frac{2}{3}$ and a beginning number, and have him or her name the next three numbers that continue the pattern.

NUMBER SYSTEMS

Expectations

Early in the year, your child will recognize and explain the relationship among 0, positive numbers, and negative numbers within a given context.

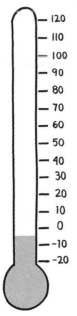

The thermometer shows a temperature of 5 degrees Fahrenheit below zero. Zero is the freezing point.

Your child will also find and plot rational numbers on a number line and use the number line to compare them.

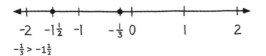

Soon after this work comparing numbers, your child will review factors and multiples and learn how to find the greatest common factor and the least common multiple for pairs of numbers.

2	24	32
2	12	16
2	6	8
	3	4

The greatest common factor of 24 and 32 is $2 * 2 * 2 = 8$.

The least common multiple of 24 and 32 is $2 * 2 * 2 * 3 * 4 = 96$.

The grid method is a useful strategy for finding both the greatest common factor and least common multiple. You can read about the grid method in your child's *Student Reference Book.*

Your child will also plot ordered pairs to draw polygons on a coordinate grid, and use absolute value to find distances between points that share a coordinate. (See also the expectations under sixth-grade "Geometry.")

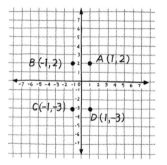

The length of *BC* is represented by $|2| + |-3|$.

By the end of grade 6, your child can be expected to find, order, and compare rational numbers on number line diagrams; understand the absolute value of a rational number as its distance from 0 on a number line; and use absolute value to find the length of polygon sides on a coordinate grid.

Do-Anytime Activities

- Have your child locate two numbers on signs or billboards and find the greatest common factor and the least common multiple of the pair of numbers.

- Play *Hidden Treasure* with your child. The directions for this game can be found in the grade 6 *Student Reference Book* or online in the Student Learning Center.
- Play a game in the kitchen or in another room of your house. Along with your child, imagine a vertical number line with the countertop as 0; everything above is referenced by a positive number and everything below is referenced by a negative number. Give your child directions for getting out items by using phrases like *the −2 mixing bowl*, which would identify the bowl on the second shelf below the counter. Something similar could be done with items on bookshelves or tool shelves.

DATA AND STATISTICS
Expectations
Early in the year your child will learn to recognize and formulate statistical questions. For example:

Question A: How many pages did you read this week?
Question B: How many pages does a typical sixth grader read in a week?

Question A is not a statistical question because it has only one answer. Question B is a statistical question because it generates a variety of answers. Your child will collect and organize data, display data on dot plots and histograms, and analyze persuasive graphs.

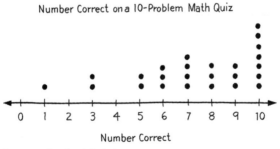

Number Correct on a 10-Problem Math Quiz

Number Correct

An example of a dot plot

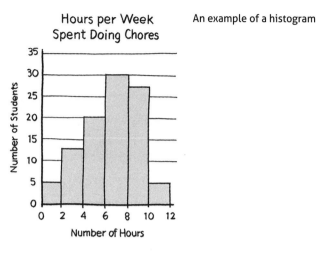

Hours per Week Spent Doing Chores

An example of a histogram

Number of Students (y-axis: 0, 5, 10, 15, 20, 25, 30, 35)

Number of Hours (x-axis: 0, 2, 4, 6, 8, 10, 12)

In addition, he or she will find measures of central tendency (mean, median, and mode) to describe the distribution of data sets and determine which one to use to summarize data sets within real-world contexts.

Student	Raffle Ticket Amount
Liam	$10.00
Amir	$4.00
Ashley	$5.00
Ben	$10.00
Camila	$8.00
Ella	$0.00
Julia	$25.00
Eric	$4.00
Brianne	$2.00
Jessica	$4.00

Mean: $7.20. Median: $4.50.
Mode: $4.00.

As the year progresses, your child will be able to display the distribution of a data set by making box plots and finding the interquartile ranges (IQR).

Time Spent Completing a 100-Piece Puzzle
(in minutes)

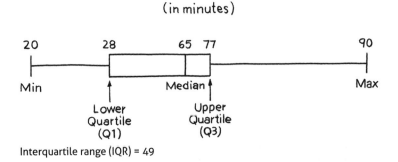

Interquartile range (IQR) = 49

He or she will then be able to calculate mean absolute deviation, or the mean of the distances of the data points from the mean, and use it to compare data sets.

Company	Beef Burger	Total Calories
Brand A	Regular Burger	290
Brand B	Large Burger	470
Brand C	Regular Burger	310
Brand D	Junior Burger	230
Brand E	Regular Burger	250

Total Calories	Deviation	Absolute Deviation
290	–20	20
470	160	160
310	0	0
230	–80	80
250	–60	60
mean = 310		mean = 64

By the end of grade 6, children can be expected to recognize statistical questions. Your child can also be expected to display numerical data in dot plots, histograms, and box plots; recognize and find measures of center, such as mean, median, and mode; and recognize and find measures of variability, such as interquartile range and mean absolute deviation.

Do-Anytime Activities

- Have your child explain the difference between a dot plot and a histogram. Ask: *What information can we gather from one that we can't gather from the other?*

- Select two different types of graphs and have your child compare them. Ask questions like these: *What types of graphs are they? How are they similar, and how are they different? What information can we interpret from the graphs? What information is not on the graphs? Can a precise mean, median, or mode be obtained from the data, or do they have to be estimated? Why?*
- Generate (or find) a list of data values and have your child create a box plot using the values.
- Ask your child to think of examples in which he or she would use the median instead of the mean or the mean instead of the median.

GEOMETRY

Expectations

Just after midyear, your child will draw polygons on a coordinate grid and use absolute value to find side lengths. (See also the expectations under sixth-grade "Number Systems.")

Your child will build geometric solids from nets. A net is a 2-dimensional figure that represents a 3-dimensional figure. It is created by cutting and unfolding or separating the figure's faces and solids. Your child will also represent prisms with nets and will use the nets to find surface area. (See also the expectations under sixth-grade "Measurement.")

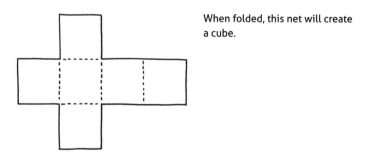

When folded, this net will create a cube.

By the end of grade 6, your child will solve real-world and mathematical problems involving nets and polygons drawn on the coordinate plane.

Do-Anytime Activities

• Ask your child to cut apart solid objects, such as cereal and tissue boxes, to create a net of the solid.

MEASUREMENT

Expectations

Just after midyear, your child will compose and decompose triangles and parallelograms to derive area formulas for parallelograms and triangles.

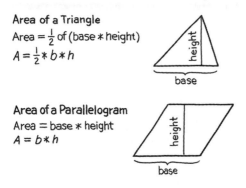

Area of a Triangle

Area = $\frac{1}{2}$ of (base * height)

$A = \frac{1}{2} * b * h$

Area of a Parallelogram

Area = base * height

$A = b * h$

Sixth graders will also find areas of more complex figures by decomposing them into rectangles, parallelograms, and triangles, finding the area of each part, and adding those areas to find the area of the larger figure.

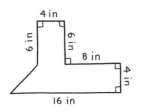

Your child will use nets to find the surface area of 3-dimensional shapes.

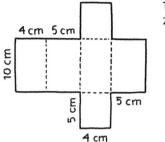

The surface area is 2(10 * 4) + 2(10 * 5) + 2(4 * 5) = 220 cm².

He or she will also find the volume of right rectangular prisms with fractional edge lengths.

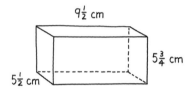

$V = l * w * h = 5\frac{1}{2}$ cm $* 5\frac{3}{4}$ cm $* 9\frac{1}{2}$ cm $= 300\frac{7}{16}$ cm³

By the end of grade 6, expect your child to solve real-world and mathematical problems involving area, surface area, and volume.

Do-Anytime Activities

- Encourage your child to estimate the area of sidewalks, backyards, patios, or different rooms in your home using a shoe as an approximate measure for 1 foot.
- Ask your child to explain how to use a net to find the surface area of a solid.
- Have your child measure and find the volume of different rectangular prisms around your home, such as package boxes, cereal boxes, or other containers.
- Give your child an area measurement for a rectangle, a triangle, or a parallelogram. Then have him or her give you as many possible measurement combinations of bases and heights as possible. For example, for a triangle with an area

of 36 cm², your child might come up with measurements of *b* = 9 and *h* = 8, *b* = 18 and *h* = 4, and so on.

- Ask your child to describe how a complex shaped area could be decomposed into more manageable rectangles and triangles to calculate its area. Have him or her sketch the area and how it could be divided into sections.

FRACTIONS

Expectations

Early in the year, your child will review fraction and mixed-number multiplication using number lines, models, and diagrams.

$\frac{3}{4} * \frac{1}{3}$

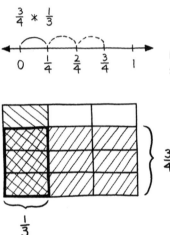

Fraction multiplication on a number line showing $\frac{3}{4} * \frac{1}{3} = \frac{1}{4}$

Fraction multiplication using an area model showing $\frac{3}{4} * \frac{1}{3} = \frac{3}{12} = \frac{1}{4}$

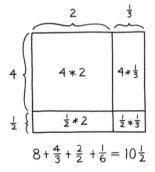

$8 + \frac{4}{3} + \frac{2}{2} + \frac{1}{6} = 10\frac{1}{2}$

Multiplication of mixed numbers using a partial-products diagram showing $4\frac{1}{2} * 2\frac{1}{3} = 10\frac{1}{2}$

He or she will also divide fractions by applying the strategies of common denominators and reciprocals.

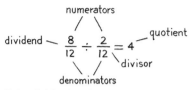

If the dividend and divisor have common denominators, and the numerator of the dividend is a multiple of the numerator of the divisor, then one can divide the numerators to find the quotient.

By the end of grade 6, your child can be expected to fluently multiply and divide fractions.

Do-Anytime Activities

• Have your child show multiplication of fractions using two different methods. Ask him or her to share how the two methods work.

DECIMALS

Expectations

After your child has some experience with fractions, he or she will use the U.S. traditional algorithms to add, subtract, multiply, and divide multidigit decimals. (For examples of U.S. traditional algorithms, see "Moving On Up . . . Operations with Bigger Numbers.")

By the end of grade 6, your child can be expected to demonstrate fluency with operations involving multidigit decimals.

Do-Anytime Activities

• Look for opportunities to perform operations with decimals. For example, have your child add the dollars-and-cents cost of two or more items. Or ask what the cost would be in dollars and cents if he or she were to buy a certain number of the same item.

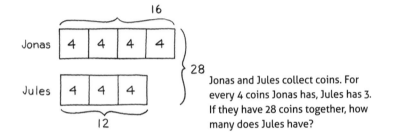

Expectations

Children begin work with ratios and proportional relationships early in grade 6. At that time, expect your child to explain the basic concept of a ratio and recognize ratio notation. A ratio is a comparison of two quantities using division. Ratios can be expressed with fractions, decimals, percents, or words. Sometimes they are written with a colon between the two numbers being compared. For example: *If Maya read 3 of the 4 books she borrowed from the library, the ratio of books read to total books can be written as $\frac{3}{4}$, 0.75, 75%, 3 to 4, or 3:4.* Your child will practice solving ratio problems using tape diagrams, pictures, and tiles.

Jonas and Jules collect coins. For every 4 coins Jonas has, Jules has 3. If they have 28 coins together, how many does Jules have?

Your child will then learn to explain what a rate is in the context of a ratio relationship. Rates are defined as ratios comparing quantities with different units. For example: *For every 2 pencils, there are 3 erasers.* Furthermore, your child will use ratio/rate tables to convert units and solve real-world problems.

Pounds	1	2	3	4	5	10
Ounces	16	32	48	64	80	160

Soon after this, your child will be introduced to percent as a rate per 100. He or she will represent a percent using a grid model and will translate between fractions, decimals, and percents.

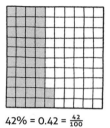

$42\% = 0.42 = \frac{42}{100}$

Your child will then learn to find the percent of a specified quantity. For example: *What is 15% of 40?* He or she will also find the whole when given a part and the percent. For example, 8 is 50% of what? For solving these percent problems, your child will use a variety of strategies and models, such as ratio/rate tables, proportional reasoning, decimal strips, equivalent fractions, and tape diagrams.

What is 15% of 40 ?

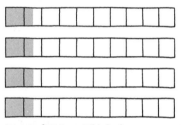

15 out of 100, or 15%, is the same as 1.5 out of 10.
Multiply 1.5 by 4 to find 6 out of 40.

As the year progresses, your child will make rate comparisons.

Car 1

miles	1,485	165	33
gallons	45	5	1

Car 2

miles	1,240	124	31
gallons	40	4	1

Which car has the better gas mileage?

By the end of grade 6, expect your child to understand and be able to explain ratio and unit rate; use ratio/rate tables to compare ratios and solve unit rate problems; use ratio reasoning to convert measurement units; and interpret and solve problems involving percent.

Do-Anytime Activities

- When purchasing items in bulk or in packages containing more than one item, have your child calculate or estimate the cost per item. For example, if you were buying a pack of 12 collectible sports cards for $1.80, you might ask your child to find the cost per card.
- Help your child find job listings in newspapers or online. If the pay is a unit rate, such as $11.25 per hour, then have him or her calculate the pay for a 40-hour week. If the pay is per month, have him or her calculate the pay for the year. If the pay is per year, estimate what the pay would be per month.
- Involve your child in event planning, for example, for birthday parties or family gatherings. Based on an estimated number of items per person or small group, have your child calculate the number of items for the total expected number of people.
- Have your child describe the relationship between two quantities using ratios. For example, at a movie theater your child might describe the ratio of children to adults, girls to total children, regular screenings to 3-D screenings, or children's movies to total movies. Ask your child whether the relationship described is a part-to-whole ratio or part-to-part ratio.
- When working with measurements, have your child convert to a different unit of measurement. For example, if you are measuring a space in feet, ask your child what the measurement would be in inches and in yards. If measuring produce in pounds, ask your child what the measurement would be in ounces.

ALGEBRA

Expectations

Beginning midyear, children evaluate expressions using the conventional order of operations. The order of operations is a set of rules that tells the order in which operations in an expression should be carried out:

1. Do operations inside grouping symbols (parenthesis, brackets, and so on) following rules 2–4. Work from the innermost set of grouping symbols outward.
2. Calculate all expressions with exponents or roots.
3. Multiply and divide in order from left to right.
4. Add and subtract in order from left to right.

For example:

$2 + 8 \div (7 - 3) * 4^2$

$2 + 8 \div 4 * 4^2$

$2 + 8 \div 4 * 16$

$2 + 2 * 16$

$2 + 32$

34

Your child will also write algebraic expressions using variables to generalize basic patterns and represent situations. For example: *Susana has n dollars. Susana has $12.50 less than Nico. How much money does Nico have?* An algebraic expression representing how much money Nico has is $n + \$12.50$.

Also around midyear, children identify equivalent expressions by recognizing expressions that represent the same value. For example, the following expressions both equal 25: $3 * 7 + 4$ and $100 \div 4$. Your child will also obtain equivalent expressions by using the Distributive Property and combining like terms. For example: $7 * (2 + 8) = (7 * 2) + (7 * 8)$. And your child will write and graph inequalities to represent constraints or conditions in real-world situations and find solution sets.

---

* Note that this differs from the order of operations shown in the algebra expectations for third grade because children at that grade level have not yet begun work with exponents or roots.

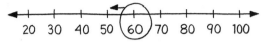

Your neighborhood yogurt shop closes when the temperature drops below 60°F.

As the year progresses, your child will write simple equations to represent situations, and use trial-and-error, bar models, pan balances, and inverse operations to solve them.

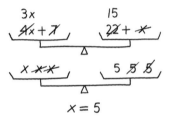

Using a bar model to solve an equation

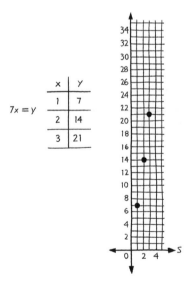

Using a pan-balance model to solve an equation

By the end of grade 6, your child can be expected to represent and solve real-world and mathematical problems with expressions, equations, and inequalities. He or she can also be expected to represent two quantities that change in relationship to one another, analyze the relationship using graphs and tables, and relate these to an equation.

Do-Anytime Activities

- Ask your child to describe patterns or situations using equations or inequalities that contain variables. For example, in reference to filling up a car with gas, your child might describe the situation in terms of the price of gas per gallon times the number of gallons needed to fill up the tank, equaling the total cost (that is, $3.05 * x$ gallons = $ y).

- Have your child come up with a mathematical expression and use it to explain how it would be solved using the conventional order of operations.

- Give your child numerical expressions and have him or her come up with equivalent expressions.

Problem Solving in *Everyday Mathematics*: How Children Think about Mathematics

We have shared throughout this book how *Everyday Mathematics* differs from other mathematics programs both in terms of the mathematical content taught and how the content is taught. In this section, you will learn how *Everyday Mathematics* also differs in the ways it teaches children to *think* about mathematics. In other words, here we will look in detail at the actual practices children engage in when they do mathematical work solving real-world problems, communicating their reasoning, and making sense of the thinking of others. Extensive research has consistently shown that children's engagement in these kinds of problem-solving activities is essential in order to truly master mathematics.

Nothing is more critical for helping children grow in their mathematical thinking than abundant experience solving true mathematics problems. While the authors of *Everyday Mathematics* understand the value of practice and include plenty of *exercises* (straightforward tasks that can be useful for practicing previously learned skills), a true mathematics *problem* differs from an *exercise*. A *problem* is a task for which children do not initially know the path to the solution; in other words, it is not something they have seen or done before. Problems are often more challenging than the more straightforward exercises. It is the abundance of true problems that *Everyday Mathematics* contains that distinguishes it from other, more traditional programs, which tend to be "problem light, exercise heavy." Maybe you've had the experience of starting on some new kind of athletic training, and you remember how sore your muscles were at first on the day or two after your workout. Over time that soreness was replaced with growing muscle, and you became stronger. The same applies to mathematical thinking. Without experiencing some strain, children are really not developing their mathematical muscles.

In order to become savvy problem solvers, children need lots of practice making sense of problems, in both real-world contexts and in purely mathematical situations. One way that *Everyday*

Mathematics engages students in daily problem solving is through the Math Message. All *Everyday Mathematics* lessons begin with a Math Message, problem for children to consider. Math Message problems are meant to be solved by children independently, in partnerships, or in small groups, and are intentionally designed to be different from any previous tasks children have encountered. Thus, each lesson begins by engaging children in genuine problem solving and mathematical thinking. As children share suggestions on how to solve this initial problem, the teacher does not immediately show children how to solve the problem. Rather he or she draws out key ideas, leading into the main content of the day's lesson.

In addition to the Math Message, there are countless other opportunities for deep mathematical thinking and problem solving throughout each and every lesson, in class discussions, during group and partner work, and as children complete independent work. Beyond this abundant daily problem solving, however, *Everyday Mathematics* includes one lesson in every unit focused specifically on deepening children's ability to understand and solve challenging mathematical problems. These lessons are called Open Response lessons and are special in the following ways:

- The lessons give children the opportunity to solve a challenging, open-ended problem. The problem can be solved in a variety of ways and may have more than one possible solution.
- The lessons are designed to take two days. On the first day, children work to solve the problem and record their thinking. On the second day, they share their thinking, interact with classmates, and make sense of each other's ideas.
- Following a class discussion on the second day, the lessons allow children the opportunity to revise their previous thinking. In other words, children don't get just one chance to solve the problem.

The two-day format of Open Response lessons makes them ideal for deeper problem solving, which often involves learning how to persevere in searching for solutions when problems are challenging. Here is an example of a third-grade Open Response task called Button Dolls:

> Tonya is making button dolls for the school fair. On each doll, she uses 2 buttons for the eyes, 1 button for the nose, and 3 buttons for the clothes. There are 8 buttons in each package she finds at the store.
>
> Tonya needs to buy packages of buttons so that all of the buttons will be used without any left over.
>
> How many packages could she buy? How many dolls will that make?

Using representations and reasoning, children learn how to apply themselves to making sense of the problem, so they can coordinate the information it contains to determine possible solutions.

$2+1+3=6$ means 6 buttons for each doll.

$$\boxed{8} \quad \boxed{6} \quad \boxed{6} \quad \boxed{6}$$

$8-2=6$ replace the 8 with 6
$2+2+2=6$ extra buttons
another 6, so we can make another doll. $\boxed{6} + \boxed{6}\,\boxed{6}\,\boxed{6} = 4$ dolls
She brought 3 packages

One third grader's solution to the Button Dolls problem

The Button Dolls problem qualifies as a true problem, instead of an exercise, because most 8-year-olds will not be able to solve it without some experimentation and serious mathematical thinking. Note also the several features of the task that make it a rich mathematics problem. It is open-ended and could lead to multiple solutions; that is, the class will probably realize that 24, 48, 72, and so on, are all potential answers to the problem, setting up an excellent opportunity for the teacher to lead a whole-class discussion of strategies and patterns and invite children to share their solutions. Finally, it is a task that is accessible to all students. That is, by using representations, most children should be able to make some progress on the task, even if they are unable to come to a solution on their own.

As children work to solve challenging problems such as the ones they encounter in Open Response lessons, they begin to learn to *persevere*, to continue trying even if the problem is difficult and they have to struggle to solve it. They are building mathematical muscles. Perseverance is essential for solving any type of difficult problem. Believe it or not, mathematicians have spent decades, if not centuries, working on particular math problems, some of which still remain unsolved today. Of course, it will not take decades to solve any of the problems in *Everyday Mathematics*, but that doesn't mean they aren't important for providing children with some productive struggle. Such hard problems help them develop the ability to persevere even as they are learning new mathematical content.

Another key aspect of problem solving that *Everyday Mathematics* encourages children to develop is *reflection*. As they work to craft solutions to the problems in *Everyday Mathematics*, the curriculum guides them to reflect on their thinking. In discussing problems with each other, children compare and consider different strategies and solution paths. Often they are prompted to consider whether the answer they have come to seems reasonable. For example, as second graders begin developing strategies for multidigit addition, they are instructed to make an estimate of the answer first. Then, after computing their actual answer, they *reflect* on their answer, comparing it to the estimate. See the bulleted list below for the prompts second graders experience repeatedly in

Everyday Mathematics; they learn not only to produce correct answers but also to reflect on their own reasoning processes so they will be able to apply and adapt that reasoning to other problems.

- Make a ballpark estimate of the answer.
- Solve the problem using any strategy you choose. Use words, numbers, or drawings to show your thinking.
- Explain how your estimate shows whether your answer makes sense.

Everyday Mathematics assignments typically include fewer items for children to solve than assignments in other math programs. This is because the *Everyday Mathematics* authors believe that it is important to given children the opportunity to solve real problems and reflect as they solve them. Doing this takes time, and we believe that time engaging in meaningful problem solving is better spent than time completing overly repetitive exercises. The opportunity to engage in solving rich problems and reflecting on the strategies used opens the door to developing deeper mathematical understanding.

Some people believe that children need to know basic facts and skills before they can do any problem solving. However, *Everyday Mathematics* believes that even the youngest learners can engage in meaningful problem-solving experiences, even before they have learned the basic facts and skills. Young children in *Everyday Mathematics* use representations, modeling, and physical tools to help them organize and share their thinking as soon as they begin solving problems. Beginning in kindergarten, children use drawings and counters (cubes, bingo chips, and so on) to represent what is happening in the number stories they explore. Older children continue to use representations to make sense of and develop more challenging mathematical ideas. For example, in third and fourth grade, they explore how representations like circle pieces and number lines can help them make sense of equivalent fractions and come to understand that the same fractional value can be expressed in different ways.

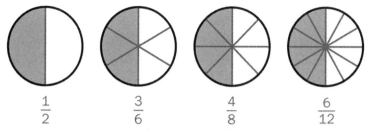

Third graders learn to see that $\frac{1}{2}$ is equivalent to $\frac{3}{6}$, $\frac{4}{8}$, and $\frac{6}{12}$ because however they partition one of the circles, the same amount of the whole is covered, just by different-sized slices. © McGraw-Hill Education.

Using tools, such as the circle pieces mentioned above, is another key component to successful problem solving. Elementary mathematics programs typically provide kits containing any number of hands-on tools for children to use, including dice, counters, shape blocks, base-10 blocks, rulers, and more. Children are expected to learn which of these tools can be helpful with certain types of calculations. *Everyday Mathematics* differs in that children learn to both *choose* and *use* such tools in the context of modeling and solving problems, and they are frequently required to choose for themselves which tool is the best one for the problem at hand. For example, a lesson from first-grade *Everyday Mathematics* poses this problem: *Damien used counters to solve 46 + 38. What other tools could he have used? Record your ideas and use one of them to solve the problem.* In the follow-up discussion, students compare various tools, such as base-10 blocks and number grids, both of which could be used to solve the problem more efficiently than Damien's counters. By wrestling with problems like this and discussing them with each other, children learn to think strategically.

The rich mathematical work of problem solving becomes even richer when children collaborate. Every *Everyday Mathematics* lesson encourages children to discuss and share their ideas with partners, small groups, and the whole class. Through this frequent experience discussing problems and problem-solving strategies, children learn how to solve problems, but, more importantly, they become able to communicate with greater precision and conviction. They learn to use proper mathematical vocabulary and to support their mathematical ideas with solid reasoning. They develop the ability to justify their thinking and to explain both orally and in writing not only what they think but why they believe what they are thinking is true. One third-grade Open Response task provides a good example. Children are asked to suppose that they are building a rectangular rabbit pen out of 24 feet of fence. They experiment with different dimensions for the pen, determining the area resulting from each option.

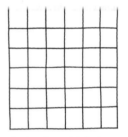

Perimeter: 24 feet
Area: 11 square feet

Perimeter: 24 feet
Area: 36 square feet
Some of the possible rabbit pens made with 24 feet of fence

When they are finished exploring the problem, children respond in writing to the following prompts:

- Which pen do you think would be best for Miguel's rabbit?
- Use mathematical language to explain the reason for your choice.

By expecting children to share both *what* they think and *why* they think it, *Everyday Mathematics* develops the expectation in children that they will be called upon to justify their mathematical thinking. Eventually this becomes such a strong habit that children offer reasons for why they think what they do without being prompted, instinctively incorporating a "because" clause following claims they share with the class.

Sharing their thinking is only half of the story. Children in *Ev-*

eryday Mathematics classrooms also develop another very important communication skill—learning to listen to others. In first- and second-grade lessons, when children are developing addition fact strategies, the teacher asks questions such as, *Does everyone understand Marsha's strategy? Can somebody explain it in your own words?* Thus, children are expected to understand their peers' ideas well enough to articulate the ideas themselves, regardless of whether they were thinking that same idea themselves. By slowing down and actually *listening* to each other, children explore ideas more deeply and keep an open mind about trying new approaches to problem solving. Throughout the program, *Everyday Mathematics* also gives children opportunities to share in writing their interpretation of others' strategies. They may be asked first to explain a strategy someone else is using in their own words and then reflect on whether the strategy seems appropriate. For example, in one third-grade task, children are presented with the written description of another child's strategy for solving 6 × 7 = ? using a helper fact:

I will use the helper fact 5 × 7. I know that 5 × 7 = 35. I can add one more group of 7 to 35 to get 42. I now have 6 groups of 7, so I know 6 × 7 = 42.

Next, they are asked to create sketches matching the strategy, as well as to explain precisely what it is about their sketches that makes them fit the child's strategy. This requires them to make sense of and represent someone else's thinking. And they reflect further on the task by identifying at least one other efficient strategy that could be used to solve 6 × 7 = ? and by writing a convincing explanation about how this alternative strategy could be used. For example, someone might propose starting with 3 × 7 = 21 and doubling the 21 to find that 6 × 7 = 42, using both pictures and words to explain their thinking. In this single task, then, children practice both aspects of communication, making sense of someone else's thinking and sharing their own.

Everyday Mathematics thus uses a unique and effective approach to encourage problem solving and mathematical communication, while most other elementary mathematics programs lack the sorts of problem-solving opportunities we have described in this section. As a result, children in traditional programs become accustomed to solving math exercises quickly while working independently. As they come to expect that success in math is supposed to be easy and fast, they lose their ability to make sense of problems, to persevere, to communicate, and to reflect on their thinking. Whenever a problem takes more than a minute to solve, they believe they either have done something wrong or are not capable of finding the solution—and quickly ask the teacher for help. Such dispositions are detrimental for future mathematical work, as children become unwilling to work on a "true problem" long enough to actually make sense of and solve it. Nor are they encouraged to develop other important life skills such as collaboration and perseverance. With deliberate attention paid to helping children not only master content, but also make sense of problems, persevere, reflect, and communicate, *Everyday Mathematics* develops more than just mathematical skill. It develops *mathematicians.*

SECTION 4
How Do I Find Out More about *Everyday Mathematics*?

At this point, we hope you have found answers to many questions you may have had about *Everyday Mathematics*. You may have a better understanding of *what* your child is learning in math, *how* he or she is learning it, and *why* he or she is learning in a way that may be so different from what you experienced in school. We also hope that what you have learned has piqued your curiosity for more. Maybe you have even more questions now than when you started. Fortunately, as an *Everyday Mathematics* parent, this book is just one of many resources available to you. In this closing section, we describe some valuable online resources that may interest you. Also note that several sections in this book include Further Reading sections, aimed at providing you with even more resources from which to choose.

The authors of *Everyday Mathematics* maintain a resource and information website at the University of Chicago that contains more information about the program: http://everydaymath .uchicago.edu. There is no access code needed. The information is free to anyone interested in *Everyday Mathematics*.

On the website, you will find a feature called *Everyday Mathematics* at Home, which provides information to assist parents for each grade level. Once you have selected your child's grade, you can select the unit and lesson that your child's class is currently studying. For each unit, you will find downloadable documents, including Background Information, Vocabulary Lists, and Do-Anytime Activities designed especially for that particular unit. In addition to brief overviews of content for the unit as a whole, the following resources are available to you:

- *Vocabulary*: vocabulary and definitions for mathematical terms that appear in the lesson

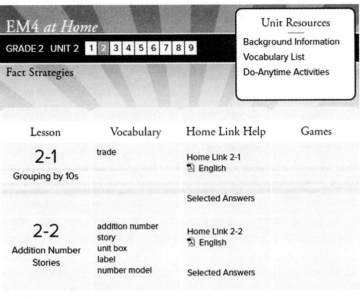

- *Home Link Help*: a downloadable Home Link for the lesson, along with selected answers
- *Games*: suggestions of games you can play with your child that correlate to the lesson, including links to game instructions in the online *My Reference Book* or *Student Reference Book* through ConnectED (when applicable).

For example, suppose you have a first grader. She comes home from school talking about "fact families," and you aren't exactly sure what she means. Worse, she forgets to bring home her Home Link. Accessing *Everyday Mathematics* at Home will provide you the help you need. Knowing she just started Unit 7, you can simply look for the lesson called "Fact Families." Clicking on the vocabulary terms corresponding with that lesson will take you to definitions to help you understand what your daughter is learning in class. Then you can download and complete the Home Link for the day. And if you are unsure whether you've interpreted everything

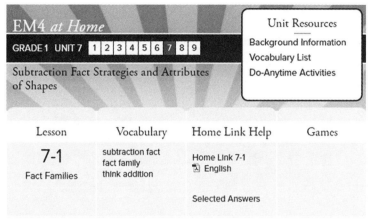

correctly, you can click on the link for selected Home Link answers to confirm that your daughter's work.

You should also have access to a web portal managed by the publisher of *Everyday Mathematics*, McGraw-Hill Education. That portal is called ConnectEd. Log on at http://connected.mcgraw-hill.com/connected/login.do with the username and password provided by your school. Once you log in, enter the Student Learning Center. From the Student Learning Center, you can access many more resources, including these:

- printable downloads for Home Links (in English and Spanish), Literature Lists, Family Letters (English and Spanish), Do-Anytime Activities, and game directions and recording sheets;
- an online, interactive version of *My Reference Book* or *Student Reference Book*;
- online versions of many *Everyday Mathematics* games;
- tutorial videos on key content ideas from the curriculum; and
- in-class assignments that were completed digitally by your child at school.

A sample screen for the Student Learning Center and the submenu for *Everyday Mathematics* at Home. © McGraw-Hill Education.

In addition to ConnectEd, McGraw-Hill Education also provides games for handheld devices (iOS and Android) modeled on the most popular *Everyday Mathematics* math games. (See "Fun and Games: Math Is So Much More than (S)Kill and Drill" for more information.)

Childhood is an incredible time of learning and growth. As a parent, you are an integral part of a partnership between home and school, whose job is to encourage the mathematical development of your child. Along with your child's teacher and the *Everyday Mathematics* program, you can play a role in shaping your child's mathematical knowledge. The resources described here from *Everyday Mathematics* are designed to enhance your ability to work with your child on math in meaningful ways. Reading this book is a great start. Hopefully, it has deepened your own understanding and appreciation for how *Everyday Mathematics* can help your child develop the understanding, skill, disposition, and confidence of a budding mathematician.

Index

adding-a-group strategy, 40
addition and subtraction: change
stories, 25–26; combinations
of 10, 35; comparisons, 27–28;
developing flexibility in working
with numbers, 33–35; doubles,
35; fact families, 37; focus on
context, 24; helper facts, 35–36;
hints for parents to help, 26, 37;
increasing the level of difficulty,
25–26; learning to connect
strategies, 37; making 10s, 35–36;
multidigit, 45–49; multistep, 28;
near-doubles, 35–36; number
stories and, 33; parts-and-total,
26–27; situation diagrams, 26;
subtraction, 37
Addition Top-It (game), 15, 96
additivity of length, 79–80
algebra: appropriateness for
kindergarten, 53–54; fifth grade,
178–79; first grade, 121–23; fourth
grade, 164; Frames and Arrows,
60–62; function machines, 62–
65; kindergarten, 109–10; modeling using number sentences
and variables, 57–59; modeling
with situation diagrams, 54–57;
patterning, 59–62; second grade,
135–37; sixth grade, 192–95; third
grade, 150–52; "What's My Rule?,"
62–64
algorithms: partial-products, 50–51;
partial-quotients, 51–52; partial-
sums addition, 47; problem
with memorizing, 74; program's

approach to, 44–45; traditional,
43–44
Area and Perimeter Game (game), 95
As You Help Your Child (Family
Letter section), 87
automaticity, 32, 38

base-10 blocks, 14
basic facts: absence of timed tests,
42; fluency and (*see* fluency);
importance of mastering, 30–31;
learning progression, 31; opportunities for practice, 31; phases
of mastery, 31–33; traditional
teaching methods, 31
Beat the Calculator (game), 90
Beat the Computer (game), 97
Bjork, Robert, 7
blocked curricula, 6
break-apart strategy, 49–50
breaking apart, 40–41
broken calculator activity, 89
Building Skills through Games
(Family Letter section), 86

calculators: activities using, 89;
development of judgment about
using, 90; game play for learning,
90; misuse of, 90–91; role in skill
development, 88–92, 89–90
change diagrams, 55, 58–59
change number stories, 25–26
circle pieces, fraction, 68–69
combinations of 10, 35
Common Core State Standards for
Mathematics, 23, 99